MICHIGAN MOLECULAR INSTITUTE
1910 WEST ST. ANDREWS ROAD
MIDLAND, MICHIGAN 48640

Plastics Product Design Engineering
HANDBOOK

Plastics Product Design Engineering HANDBOOK

Sidney Levy
J. Harry DuBois

MICHIGAN MOLECULAR INSTITUTE
1910 WEST ST. ANDREWS ROAD
MIDLAND, MICHIGAN 48640

MIDLAND MACROMOLECULAR INSTITUTE
1910 WEST ST. ANDREWS DRIVE
MIDLAND, MICHIGAN 48640

VAN NOSTRAND REINHOLD COMPANY
NEW YORK   CINCINNATI   ATLANTA   DALLAS   SAN FRANCISCO
LONDON   TORONTO   MELBOURNE

Van Nostrand Reinhold Company Regional Offices:
New York   Cincinnati   Atlanta   Dallas   San Francisco

Van Nostrand Reinhold Company International Offices:
London   Toronto   Melbourne

Copyright © 1977 by Litton Educational Publishing, Inc.

Library of Congress Catalog Card Number: 77-1044
ISBN: 0-442-24764-8

Manufactured in the United States of America

Published by Van Nostrand Reinhold Company
450 West 33rd Street, New York, N.Y. 10001

Published simultaneously in Canada by Van Nostrand Reinhold Ltd.

15 14 13 12 11 10 9 8 7 6 5 4 3 2 1

Library of Congress Cataloging in Publication Data

Levy, Sidney, 1923–
    Plastics product design engineering handbook.

    1. Plastics.   2. Engineering design.
I. DuBois, John Henry, 1903–   joint author.
II. Title.
TP1122.L45   668.4'9   77-1044
ISBN 0-442-24764-8

Preface

This is the first book directed to the solution of the problems of the plastics product designer that treats the subject from fundamentals to specifics with the applicable mathematics. Material economy developments necessitate a most rigorous approach to product design since materials costs have risen greatly and the user performance expectations have enlarged tremendously.

Plastics are materials that behave in a complex manner when subjected to physical and environmental stresses. The intuitive design approaches used by many designers to overcome the lack of knowledge of the performance characteristics of these materials has lead to a combination of over-design and product failures. The approach here is to review the fundamentals of material structures and provide a background for an analytic design procedure for plastics products. This is carried out in the area of structural loads, cyclical loads, electrical field exposure, environmental exposure, and the other types of in-use exposure to which plastics components may be exposed.

Using the approach developed from the fundamentals, the text describes the design of plastics products in a variety of different fields and illustrates this with specific examples with solutions. Some examples are computer assisted to illustrate the method of arriving at a safe, effective, and economical design. The interplay of the various elements in the design such as aesthetics, customer acceptance, environmental influences, and the effects of the manufacturing processes on design and product performance is examined and a procedure given for making trade-offs for maximum utility and economy.

The need and use of product and material testing as part of the design process is covered. The relationship between the level of risk involved in the product use and the testing philosophy is explored so than an economical level of testing is arrived at consistent with successful usage. The role of the design engineer in the total product development process is also part of the discussion since it affects the specific responsibility of the designer in his job.

The authors believe that the text will aid in the design process and in the training of design engineers to work with plastics. Plastics are the most versatile of the engineering materials and we feel that a better understanding of the strength and the limitations of the materials will lead to highly successful and economical applications for the plastics.

Sidney Levy, P.E.
J. Harry DuBois

Introduction

Plastics have become increasingly important in the products used in our society, ranging from housing to packaging, transportation, business machines and especially in medicine and health products. Designing plastic parts for this wide range of uses has become a major activity for designers, architects, engineers, and others who are concerned with product development.

Because plastics are unique materials with a broad range of properties they are adaptable to a variety of uses. The uniqueness of plastics stems from their physical characteristics which are as different from metals, glasses, and ceramics as these materials are different from each other. One major concern is the design of structures to take loads. Metals as well as the other materials are assumed to respond elastically and to completely recover their original shape after the load is removed. Based on this simple fact, extensive literature on applied mechanics of materials has been developed to enable designers to predict accurately the performance of structures under load. Many engineers depend on such texts as Timoshenko's *Strength of Materials* as a guide to the performance of structures. Using this as a guide, generations of engineers have designed economical and safe structural parts. Unfortunately, these design principles must be modified when designing with plastics since they do not respond elastically to stress and undergo permanent deformation with sustained loading. Plastics also have a low stiffness compared with other materials; they cover a range of stiffness of six orders of magnitude and exhibit a high dependency of physical properties on temperature. It is obvious therefore that the structural design of plastics parts requires a different design approach.

This is also true in other areas such as thermal design, environmental design, and geometric design of parts where the unique properties of plastics require a different design approach. This book is intended to show how plastics are different from other engineering materials, and to develop the relationships necessary to design intelligently with these materials. The book proceeds from a basic understanding of how the physical structure of these materials differs from other materials and describes how these differences affect their performance under various stresses. It also describes how to use this information in the design of products.

The results of processing the plastic materials affect their performance profoundly. The basic information of plastics structures is given to enable the designer to predict the performance of his product within the limitations of the processing. The unique effects of the environment on plastics are examined so that the designer can design a product for a lifetime exposed to the "use" environment.

The text is divided into four sections. The early chapters cover the physical structure and molecular configuration of the polymer materials which are the basic materials of plastics. It shows how the microstructure of the materials leads to the observed physical and chemical properties and develops the basic relationships between the response of the materials and the applied stresses.

The next section is concerned with developing the mathematical relationships and formulas to describe the response of plastics members to static and dynamic loads. Specific design situations are analyzed and the methods of problem solving developed. Because of the complex nature of the materials, solutions are frequently not possible. The use of graphical solutions, curve-fitting techniques, and computer assisted design techniques are explored.

The section that follows covers the geometrical considerations in design, with particular emphasis on the methods of improving the effective stiffness of plastics structures and of compensating for the nonlinear stress-strain behavior of the materials. The manufacturing processes impose additional restrictions on design and the way in which these affect the design process is described. The processes for manufacture allow freedom in design as well. They permit the extension of functional design concepts to a degree not possible with other materials by the combination of functions in one part which, in other materials, might require several parts.

The next section covers the effect of the end-use environment on product performance so as to permit the design of a product for a reasonable lifetime with realistic economics. Included is the design engineer's involvement in testing and evaluating, as well as a definition of the specific function of the part, to prevent its abuse. The involvement of the product designer is such that his responsibility extends to having the product serve the end-user effectively, and the text indicates the extent and means whereby he can do this.

This book is written at a time when the nature of plastics products and the applications of plastics have been drastically altered. Plastics are no longer inexpensive materials to be substituted for other materials with little regard for their efficient useage. The energy and materials shortages in the raw materials to make plastics preclude any such possibility. Plastics products must be efficiently designed using rational methods of determining part sizes to meet the anticipated loads and other stresses. In addition, the environment of the design process has changed. The legal strictures on design for safe use, the environmental pressure caused by the existence of large numbers of plastics products such as packages, and the direct legal responsibility of the designer for the safety and utility of a product creates additional problems for the plastics product designer. It is hoped that this text will provide a tool for the efficient use of plastics materials in the design of safe, economical products. The use of functional design concepts that employ the unique properties of plastics materials can make possible a large variety of new and successful areas of application. The concluding chapter projects some of these areas.

This book is the outgrowth of lecture courses in plastics product design, a number of articles on specific areas of plastics engineering design, many years of combined experience of the authors in bringing plastics products to the market, and thirty years experience with texts on plastics and the editing of journals. The authors hope that it will be helpful to designers in utilizing plastics and polymer-based materials, those complex and interesting materials that nature used long before man to create the interesting and beautiful forms of nature, including man himself.

Acknowledgments

The authors gratefully acknowledge the assistance and encouragement of Professor Herman Mark, Professor A. G. H. Dietz and Harry S. Katz for their helpful suggestions and review of the manuscript. Permission to reproduce data from many important texts by leading authors is greatly appreciated. We also thank the many material and machinery makers who supplied illustrations for this book. Charlotte A. Levy has been a tremendous help in typing and preparing the copy.

Sidney Levy
Box 87
Thorofare, N.J. 08086

J. Harry DuBois
Box 346
Morris Plains, N.J. 07950

Contents

Plastics Product Design Engineering HANDBOOK

Chapter 1

Polymer Structure and Physical Properties

Each material that we come in contact with has its own special properties. Some are hard while others are soft. Some things are heavy and others are light. Some materials feel warm while others feel cold. These subjective properties are related to the intrinsic physical properties of material which can be defined and measured on an appropriate scale. Some of these properties are listed in Table 1-1.

The properties are related to the atomic and molecular structure of the materials. For example, we could compare iron, glass, salt, and polyethylene plastics. They represent a metallic crystalline solid, a vitrified amorphous ceramic, a nonmetallic crystalline solid, and an amorphous high polymer. Figures 1-1, 1-2, 1-3, and 1-4 show the structures in schematic form for each of these materials.

Metallic crystalline solids such as iron, nickel, or aluminum are hard stable materials. They conduct electricity and heat. They range in density from 2 g/cc up to the most dense solids known. Objects made from metallic crystalline materials respond to forces or loads applied in an elastic manner, that is, they deform in proportion to the applied force and recover to their original dimension when the distorting force is removed. When the applied force exceeds a characteristic value, the elastic limit, the material flows and the part deforms permanently. After the deformation, however, the part still behaves elastically at loads below the elastic limit.

The ductility, the extent of permanent deformation possible, varies widely between different metals and when the ductile limit is exceeded, the material becomes brittle and breaks.

Glassy materials exhibit widely different properties. They conduct heat well but are electrical nonconductors. They are generally transparent to light and other electromagnetic radiation. Their initial

1

Table 1-1. Physical Properties of Various Materials.

	Steel	Copper	Polyethylene	Phenolic	Glass
Specific gravity	7.8	8.9	0.92	1.3	2.5
Thermal conductivity					
(Btu/hr/sq ft/°F/in.)	27	200	0.19	0.3	0.7
Thermal expansion					
(in./in./°C)	8.2	9.8	120	15	5.1
Specific heat	0.1	0.1	0.5	0.4	0.2
Modulus of elasticity	30×10^6	17×10^6	0.15×10^6	10^6	10×10^6
Hardness (Rockwell)	C40	C15	R50	M110	C90
Electrical resistivity	19	2.03	10^{23}	10^{17}	10^{21}

response to applied load is an elastic deformation with complete recovery if the load is applied for relatively short periods of time (hours or days). Under sustained load, they deform permanently even at loads well below the elastic limit. The rate of cold flow or creep is temperature and time dependent as with most materials.

Glassy materials are brittle; they can be deformed elastically by small amounts, generally under 1% and they fracture when the limit is exceeded. They are very sensitive to shock (high rate loading) and are generally considered poor structural materials despite their high strength. While the physical properties of metals are affected by the processes used in their manufacture, glassy materials are much more

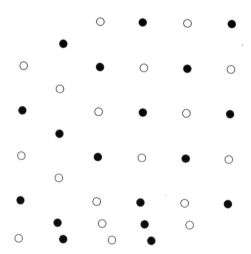

Fig. 1-1. General crystalline structure of an ionic solid-like substance.

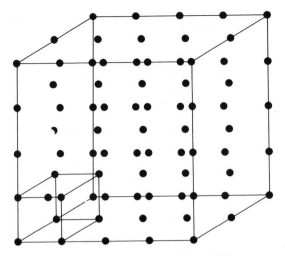

Fig. 1-2. Crystalline structure of a metallic item.

sensitive to processing effects. Stresses induced in glass objects during working can cause sudden failure even without external loading. These stresses are induced during cooling from the melt.

Nonmetallic (ionic) crystalline solids such as salt are nonconductors of electricity and fair conductors of heat. They vary widely in

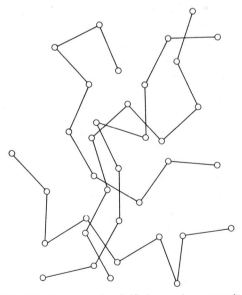

Fig. 1-3. Structure of a vitrified amorphous ceramic.

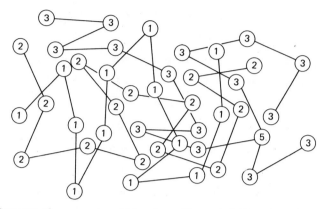

Fig. 1-4. Structure of an amorphous high polymer. Numerals facilitate following the individual chains.

density but are generally in the less extreme ranges from about 1.0–5.0. They differ from the metallic crystals in lacking free electrons which provide both electrical conductivity and high thermal conductivity. The crystalline solids deform elastically under applied forces and, depending on the type of crystal structure, can undergo brittle fracture or ductile deformation at force levels above the elaslic limit. They generally are hard materials with a high modulus of elasticity (force per unit of deformation). The ionic crystals are frequently transparent and often are colored. Polycrystalline ceramics such as densified alumina are strong materials although they do undergo brittle failure. They are frequently capable of absorbing large (high rate) shock loads. They exhibit a low order of deformation under sustained load and are among the most stable materials known, even at high temperatures.

Polymeric (plastics) materials such as polyethylene have a wide range of physical properties. Some are transparent and most pass electromagnetic radiation over a wide spectral range. The range of elastic modulus for plastics unreinforced by other materials is low, ranging from slightly over 1,000,000 psi down to 10 psi. Plastics are nonconductors of electricity and poor-to-fair conductors of heat. The structure of polymers combines some glassy structure, some nonmetallic crystalline structure, and some amorphous structure and their response to applied forces is quite complex. For short duration (1/1000 seconds) loading, many plastics respond elastically or viscoelastically. This elastic response is damped, and the deformation and recovery is not instantaneous: this viscoelastic behavior is common to all polymer-based materials. Longer loading times lead to permanent creep or cold flow deformation for most plastics. Deformation occurs at relatively low force levels and becomes more pronounced at higher force levels. The effects are highly dependent on temperature increasing in extent at elevated temperatures.

The properties are related to the atomic and molecular structures of the materials. Figure 1-5 shows an ionic crystalline solid (metal or nonmetal) and the response to an applied force. The type of force is a shear stress which is the easiest to visualize and can be related to the tensile and compressive loads normally used. The atoms of the solid are held together by the electrostatic attraction of the atoms for each other. This force is strong and the deformation, by pushing the planes of the atoms, results in increasing the distance between the atoms which requires work. When the deforming force is removed, the electrostatic fields return the atoms to the equilibrium position which shows as complete recovery. The times involved are of the order of 10^{-10} to 10^{-12} second which is, in effect, instantaneous. When the applied force is sufficiently high, the planes of atoms can move past each other to assume the next equilibrium set of positions. At this point the material has passed the elastic limit and is undergoing permanent deformation. In the case of ductile materials, the planes slip past each other and there is flow of the material. In the case of brittle materials, the material will cleave on the planes where the slippage occurred and split. With ductile materials, as the planes continue to slip past each other, the material undergoes a recrystallization process in which the energy of the material is increased and it becomes progressively more difficult to deform the material. This effect is referred to as *work hardening*, and eventually the material becomes brittle and fails by fracture. The actual details of the processes for

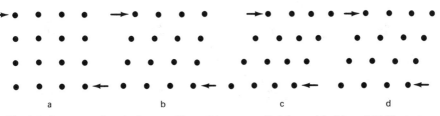

Fig. 1-5. Response of an ionic crystalline solid to an applied force. (a), (b), and (c) illustrate elastic displacement. (d) is a plastics permanent displacement.

different materials are quite complex, but the principles just described apply in all cases.

The restoring forces in the ionic solids are direct electrostatic fields. In covalently bonded materials the electron transfer results in a different configuration of the restoring forces. Usually the covalent bonding in ceramics and glasses fixes the angle between the atoms that make up the structures (Fig. 1-6). The stresses applied to the material have the effect of distorting the bond angle which in turn changes the electric fields holding the bond angles. Removal of the external forces permits the bond angles to be restored and this restoring force is also completely elastic with typical restoring times of 10^{-10} or less seconds.

The glassy structures consist of covalently bonded chains of simpler molecular elements which are amorphously distributed among each other in a dense packing arrangement. Short-term stresses bend the covalent bonds so that the materials behave elastically. Figure 1-7 shows how the chains can slide past each other with a minimum of energy required so that with stresses applied for extended periods of time the material can flow or creep. The fracture mechanism usually involves separation of the chains from each other and this results in the typical shiny concoidal fracture exhibited by glassy materials.

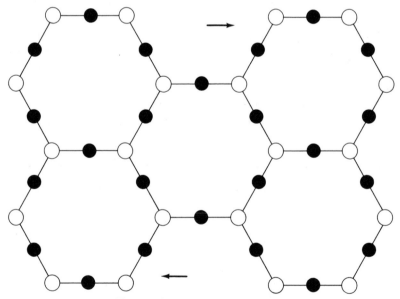

Fig. 1-6. Covalently bonded plastics.

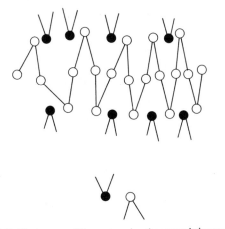

Fig. 1-7. Chains can slide past each other permitting creep.

Polymers are long chains of simple chemical units referred to as *monomers*. The structure of plastics and the behavior of the materials are based both on the chemical nature of the individual monomer link and the way in which the chains fit together. In most polymer materials the covalent bond between the individual monomer units has freedom of rotation so that the bond angle may be fixed while still permitting enough rotation to allow the chains to alter their configurations from coiled structures to stretched chains. Figure 1-8 shows some typical configurations of polymer chains including some where there is sufficient polar attraction between parts of the molecule to re-

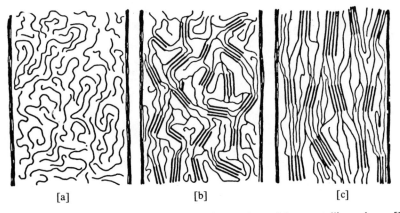

[a]   [b]   [c]

Fig. 1-8. Schematic representation of an amorphous polymer [a], a crystalline polymer [b], and an oriented crystalline polymer [c]. (From *Mechanical Properties of High Polymers*, Turner Alfrey, Jr. Interscience, 1948)

sult in crystalline areas which react as you would expect a crystalline material to react.

When a plastics (polymer) is subjected to stress, the structure can react in a number of ways. In the first reaction the bonds are stressed by stretching or bending which is the *elastic response*. Unlike the more ordered structures, adjustments of the strain in the materials is hindered by the interference between molecules so that all but the very initial response is hindered by frictional effects and the material shows a delay between the application of the stress and the resulting strain. This behavior is referred to as *viscoelastic behavior*. From Figure 1-9 it can be seen how the molecules slide past each other to increase spacings and reduce the elastic load on the bonds. Sustained stress causes actual displacement of the molecular chains with extensive movement of the chains past each other and results in flow-like behavior which is referred to as *creep* or *cold flow*. At constant initial strain the slippage of the molecules and the adjustment of position lead to another condition which is called *stress relaxation*. The level of the resistance of the structure to applied deformation drops and the material assumes a lower energy configuration.

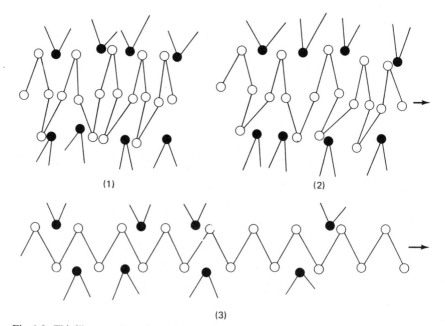

Fig. 1-9. This illustrates how the molecules slide past each other, increasing their spacing when a stress is applied to polymeric solid.

High polymers, which are the structural elements in plastics materials, can have a wide range of configurations, by the physical and chemical nature of the repeating monomer element and in part by the phase changes in processing. The simplest materials such as polyethylene exist in an amorphous state with a molecular level appearance which has been likened to a bowl of spaghetti. This is a useful picture with one modification shown in Fig. 1-10, namely, that it is kinky spaghetti with a bend between each repeating unit. There is little or no polar attraction between the chains and only molecular attraction forces of the Van der Wahl type hold the chains together. The material is readily deformed, has a low elastic modulus, and is subject to continuing deformation under load or creep at relatively low loads (Fig. 1-11).

The amorphous state in polyethylene is the natural one resulting from processing where the material is quenched from the melt and can be readily recognized because the material is transparent with relatively low haze levels. Nothing in nature is simple. Even polyethylene, when slowly cooled from the melt, produces regions where the orderly chain length causes crystallization of parts of the material by folding the chains back and forth on itself, as well as by the stacking of adjacent chain segments. These regions of orderly arrangement, which are different from the spaghetti, drastically alter the behavior of the material. When the chains are crystallized together, the apparent effect is crosslinking the material. The crystalline areas respond much more elastically to stress so that the overall stress-strain behavior moves closer to an elastic response. Even the appearance of the material changes, with a milky opaque appearance instead of the clear transparent condition. The crystalline level affects the tensile

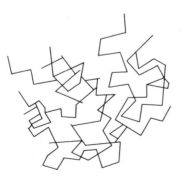

Fig. 1-10. Polyethylene exists in an amorphous state with a kinky molecular level appearance.

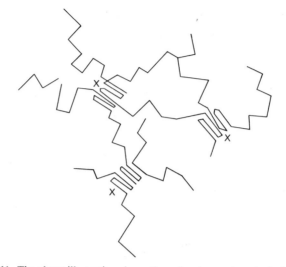

Fig. 1-11. The above illustration shows the ordered areas in polyethylene (x).

strength, the modulus of elasticity, the impact strength and other properties, making polyethylene a generally stronger material.

There are three varieties of polyethylene that are recognized commercially, namely, conventional or low density polyethylene, linear or high density polyethylene, and ultra-high molecular weight polyethylene. The main difference in structure of the two principal varieties is that the low density material has a highly branched structure while the linear or high density material has long straight chains with little or no branching. It is obvious from the name that one resulting difference is the density. The long straight chains pack together more closely, occupying less volume and resulting in higher density. In addition, the more orderly structure encourages crystallization of portions of the material so that high density polyethylene is generally much more crystalline than low density polyethylene. The result is a much stiffer, higher strength material, generally milky opaque with better temperature resistance than low density polyethylene.

Before examining some other typical polymer structures, two other interesting structural arrangements which have drastic effects on polymer properties are worth mentioning. Polyethylene plastics, although they are thermoplastic materials capable of repeated remelting and reforming, can be crosslinked. Bonds can be formed between the chains by chemical or physical processing. The result of

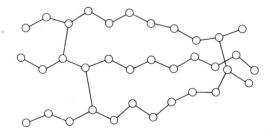

Fig. 1-12. Bonds may form between the chains resulting from chemical or physical action producing crosslinked polyethylene.

the crosslinking is shown in Fig. 1-12. The ability of the chains to move past each other is severely restricted; the more restricted, the higher the degree of crosslinking. This results in several changes in the properties of the material. The material increases in temperature resistance to flow and increases in stiffness and strength. At very high levels of crosslinking, the material becomes hard and brittle and is a thermoset plastics. Other properties such as the electrical and surface properties are also changed (See Table 1-2).

The other structural change involves the orientation of the material. If a sheet or strip of polyethylene is raised to a temperature called the *orientation temperature* and stretched either in one direction or biaxially, the physical properties are drastically altered. The stretching actions cause the molecules to slide past each other and change from a random spaghetti configuration to one in which the molecules are lined up as the spaghetti was in the package. If a stress is applied to the material in the orientation direction there is little or no chain bending and slipping. The bonds are directly stressed and the material

Table 1-2. Comparison of the Physical Properties of Polyethylene and crosslinked Polyethylene.

Properties	Medium Density Polyethylene	Polyethylene Crosslinked Compounds, Molding Grades
Tensile strength, psi	1200–3500	1600–4600
Elongation, %	50.0–600.0	10.0–440.0
Tensile elastic modulus, 10^5 psi	0.25–0.55	0.5–5.0
Thermal conductivity, 10^{-4} cal/sec/ sq cm/ 1(°C/cm)	8.0–10.0	
Thermal expansion; 10^{-5} in./in./°C	14.0–16.0	10.0–35.0

responds in a highly elastic manner compared with the unoriented material, with much higher tensile strength and elastic modulus. The elongation percentage drops and the hardness is increased (See Figs. 1-13 and 1-14).

There is a decrease in the relative physical properties across the orientation direction, although they are at least as good as for the unoriented materials.

Polystyrene is an example of a polymer which has a somewhat more complicated structure than polyethylene as shown in Fig. 1-15. The benzene ring attached to the essentially polymethylene structure adds a restriction on molecular movement by steric hindrance, or more simply put, by interference of the pendant phenyl group slipping past other such groups. Polystyrene, as might be anticipated, is a more rigid material than polyethylene and is glossy in nature at room

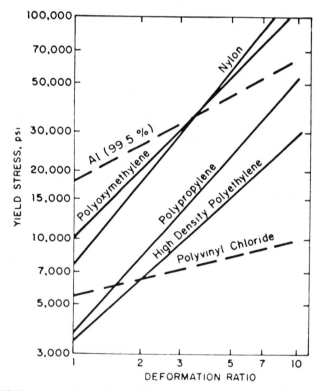

Fig. 1-13. Yield stress vs deformation ratio for various materials. (From *Polymer Engineering and Science,* **14** (10) Oct. 1974)

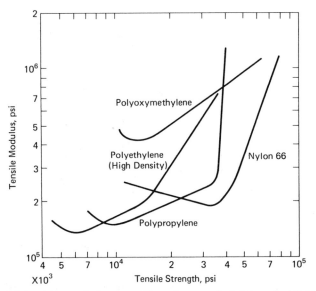

Fig. 1-14. Modulus vs strength for crystalline polymers oriented by uniaxial rolling. (From *Polymer Engineering and Science*, **14** (10) Oct. 1974)

temperature, a condition not necessarily anticipatable from the structure. There is a great deal of resistance to chain interaction and, consequently, polystyrene has no tendency to form layered structures and does not crystallize. At elevated temperatures, the material softens to a highly viscous melt and in between it exists as a highly damped elastomer or rubbery solid.

The stiffness of the polystyrene is caused by the steric effects and also by a different type of attraction between the polymer chains than polyethylene. The phenyl structure is somewhat polar (forms an electric dipole as shown in Fig. 1-16), and there is a low level of

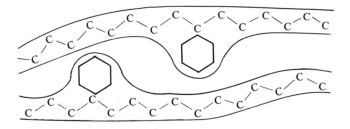

Fig. 1-15. Steric effect of bulky groups on the packing of polymer molecules.

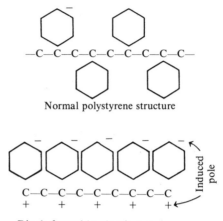

Fig. 1-16. The field effect under electrical stress forms an induced dipole in polystyrene.

electrostatic attraction between the polymer sections and adjacent chains. As a result of this combination the material reacts to stress more elastically since bond bending, an elastic mechanism, is prevalent because of the restricted movement of the chains. Table 1-3 shows some of the properties of this material. Although polystyrene does not crystallize, it can be crosslinked and oriented with effects similar to those shown by polyethylene.

Polyvinyl chloride polymer has a different pendant group on the polymethylene chain which introduces another type of chain interaction and a different set of properties and property variations. The chlorine atom is a highly polar addition to the polymer chain and produces a large electric dipole into the polymer structure. There is

Table 1-3. Physical Properties of Polystyrene.

Physical Properties	Polystyrene
Tensile strength, psi	5000–12000
Elongation, %	1.0–2.5
Tensile elastic modulus, 10^5 psi	4.0–6.0
Compressive strength, psi	11500–16000
Impact strength, ft lb/in. of notch	0.25–0.40 (1/4 in. bar) 1.6 (1/8 in. × 1/2 in.)
Thermal conductivity, 10^{-4} cal/sec/sq cm/ l(°C/cm)	2.4–3.3
Specific heat, cal/°C/gm	0.32
Thermal expansion, 10^{-5} in./in./°C	6.0–8.0

a strong attraction of the chains to each other and the resulting polymer is a rigid material. At room temperature vinyl homopolymers are brittle, hard resins which exhibit fairly elastic behavior because of the extensive dipole interaction. At elevated temperatures, the vinyl resin materials exhibit elastomeric properties (i.e., rubberlike) for two major reasons. The dipole interaction along the chains causes the polymer chains to coil and extension of these coiled chains acts to exhibit the typical elastomer interaction. The interchain dipole action acts like a virtual crosslinking action to enchance the effect.

Polyvinyl chloride polymers are commonly made into compounds by the addition of plasticizers, which are liquids that have high boiling points and dissolve the resin at elevated temperatures. The plasticizers are attracted to the polar portions of the chain and have the same effect as elevated temperature on the polymer. The softening point is lowered and the plasticized PVC compounds are elastomers at room temperature. The strong chain interaction and the viscous nature of the plasticizers cause the response to be damped so that the deformation and the recovery are slowed. This makes the materials interesting substances for use in vibration damping, but limits true elastomer applications.

PVC polymers can be oriented and crosslinked in a manner similar to the other resins discussed. Because of the polar side grouping attraction, the properties of the material are improved in both the orientation direction and in the transverse direction. The orientation has the effect of moving the polymer chains closer together so that the strong dipoles on the polymer are moved closer together to enhance the interchain attraction. PVC introduces one other characteristic in plastics materials which is important in the application of the material. PVC materials have a strong plastic memory. If the material is deformed at slightly elevated temperatures it will not recover immediately. At some later time, if heated to a temperature near the heat distortion temperature, the material "remembers" its original shape and returns to that shape. This property is used in some forming processes, but it can be a problem in improperly molded parts, causing unmolding, a reversion to the form of the material before the part is formed.

Polymers like the polyamides (nylon) and the polyesters are more polar materials and exhibit a high degree of crystallinity. There are substantial areas of the polymer structure where both chain folded and interchain crystallization takes place. Both polymers are higher melting materials having higher heat distortion temperatures and generally higher strength and more elastic properties because of the extensive crystalline areas. The crystallized regions behave like elastic crystals and the strong interchain attraction between the polar groups on the polymer chains produces generally good elastic response. Strongly oriented nylons and polyester materials are among the strongest polymer materials known with tensile strengths in the range of 50,000–200,000 psi and elastic moduli up to 1,000,000 psi and higher. These materials are used to make fiber, high strength films and similar materials where the orientation produces chain alignment as well as extensive crystallization with a resulting double strength enhancement.

The discussion so far has been limited to thermoplastic materials which are not intended to change chemically during processing and which are, in theory, capable of repeated remelting and reforming. Thermosetting polymers are chemically active materials that react during the molding or other forming process. They contain double bonds or other structures which have the effect of crosslinking the polymer chains during the forming process. The density of crosslinks is usually quite high so that a network structure is formed (Fig. 1-17). The crosslinks strongly inhibit the deformation of the polymer structure since the chains have limited mobility before a bond is stretched and must be ruptured to produce further motion. As a result, the properties of a typical thermoset are substantially different from a similar thermoplastic; for example, compare the properties of a thermoplastic and thermoset polyester.

The thermoplastic polyester gains most of its strength from the stiff chains in the polymer plus the strong polar interaction between the

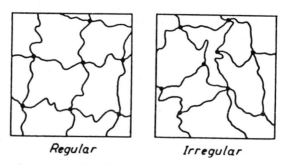

Regular   Irregular

Fig. 1-17. Schematic representation of regular and irregular network structures. (From *Mechanical Behavior of High Polymers*, Turner Alfrey, Jr., Interscience, 1948)

molecules. The thermoset polyester is generally not as polar because it has a substantial unsaturated vinyl structure for crosslinking and, after the crosslinking has taken place, the primary strength is in the interchain bonding. The thermoset polyester does not soften to a melt at elevated temperature and is capable of resisting higher operating temperatures without deforming. The thermoset polyester is generally stiffer, has a higher elastic modulus and a higher tensile strength. Because the materials cannot deform to any great extent without bond rupture, the materials tend to be brittle and have low impact strength. In most polyester thermoset formulations, the low impact strength is improved by the addition of fibrous fillers. This does nothing, however, to improve the low elongation to failure of the materials which is characteristic of heavily crosslinked polymers.

Thermoset polyesters are typical of the thermoset plastics—the characteristics of which are good temperature resistance, good tensile strength and stiffness, a high degree of hardness, low impact strength unmodified, low extensibility, and generally superior resistance to creep and cold flow. Properties of some typical materials are shown in Table 1-4.

There is one other structural type among polymers that is important, although the polymers are not widely used because of their cost and difficulty of processing. These materials are referred to as the *ladder polymers* since the monomer units are tied together with two separated bonds that cause the materials to appear as a ladder on a molecular scale (Fig. 1-18). It is interesting to speculate, based on the previous discussion, as to what effect this restriction might have on the physical properties of the materials. Because of the double coupling between the elements of polymer chain the rotation of the bonds is completely inhibited. This rigid chain is structurally stiffer than either the linear thermoplastic or the crosslinked thermoset materials. The restriction of chain rotation has the effect of much more severely restricting relative chain motion than any of the other effects discussed. The

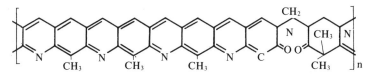

Fig. 1-18. Ladder polymers result when the monomer units are tied together with two separated bonds. (From C.S. Marvel—*SPE Transactions*, Jan., 1965)

Table 1-4. Polyesters—Cast.

Type	ASTM	Allyl Type	Rigid Styrene Type	Nonrigid Styrene Type
Physical Properties	ASTM			
Specific gravity	D792	1.30–1.45	1.12–1.46	1.06–1.25
Thermal conductivity, Btu/ hr/sq ft/°F/ft	C177	0.116–0.121	0.10–0.12	—
Coefficient of thermal expansion, per °F	D696	$2.8–5.6 \times 10^{-5}$	$3.9–5.6 \times 10^{-5}$	—
Refractive Index	D542	1.50–1.58	1.53–1.58	1.50–1.57
Specific heat, Btu/lb/°F		0.26–0.55	0.30–0.55	—
Water absorption (24 hr), %	D570	0.03–1.0	0.15–0.60	0.40–2.5
Mechanical Properties				
Mod of Elast in Tension, psi	D638	$2–3 \times 10^5$	$1.5–6.5 \times 10^5$	—
Tensile strength, 1000 psi	D638	4.5–7	4–10	0.9–1.9
Elongation (in 2 in.),%	D638	—	<5	40–310
Hardness (Rockwell)	D785	M92–118	M65–115	—
Impact strength (Izod notched), ft-lb/in.	D256	0.18–0.32	0.18–0.40	>7.0
Modulus of elasticity in flexure, psi		$3–8 \times 10^5$	$3–9 \times 10^5$	$0.001–0.1 \times 10^5$
Flexural strength, 1000 psi	D790	6–14	7–19	—
Compressive strength, 1000 psi	D695	20–26	12–37	—
Electrical Properties				
Volume resistivity, ohm-cm	D257	$>10^{13}$	$>10^{13}$	$>10^{12}$
Dielectric strength (short time), v/mil	D149	330–500	340–570	220–400
Dielectric constant				
60 cycles	D150	3.2–5.2	2.8–4.4	4.2–7.0
10^6 cycles	D150	3.3–4.8	2.8–4.0	3.7–6.1
Dissipation factor				
60 cycles	D150	0.006–0.02	0.003–0.04	0.02–0.18
10^6 cycles	D150	0.024–0.045	0.006–0.04	0.02–0.06
Arc resistance, sec	D495	120–150	115–135	125–145
Heat Resistance				
Max rec svc temp, F	D648	300	250–300	200–250
Heat distortion temperature, °F		120–320	120–420	250
Chemical Resistance		Attacked by oxidizing acids. Slightly attacked by strong alkalis. Resistant to organic solvents	Slightly to heavily attacked by strong acids. Attacked by strong alkalis, ketones and chlorinated solvents	
Uses			Castings used for aircraft glazing, electrical components, decorative applications. Resins used for premix and prepreg molding materials, matched metal molding and hand lay-up molding.	

From *Materials in Design Engineering*, Material Selector Issue, Oct. 1963.

ladder polymer materials have high elastic moduli approaching those of the metallic materials, and the crystalline versions of the ladder polymers actually have unreinforced moduli up to 5×10^6 psi.

The emphasis to this point has been on the response of polymer materials to mechanical stresses. There are several other important external effects that are important to the behavior of plastic parts. The effects of heat, electrical fields, and electromagnetic radiation are frequently important in the design of plastic parts. Since each of these effects will clarify the way in which polymers behave, they will be briefly covered as background for the design relationships to be developed in the following chapters.

The effect of heat on polymer materials is to soften them and eventually to cause them to form a viscous melt in the case of thermoplastic materials, or to form a gel-like soft structure in the case of thermoset materials. Other effects of heat such as degradation and thermally induced crosslinking may obscure these effects in some materials but, in general, these states are observable for all polymers. There are several molecular level phenomena which cause the softening. First, the additional molecular excitation caused by the heating causes the molecular chains to increase their average distance from each other. The density is reduced, but from a physical stress response standpoint, the most significant effect is that the interchain reactions are reduced so that the materials deform much more easily at lower stress levels. The elastic response by bond bending occurs at lower levels and is not as hindered by steric effects, and the viscoelastic response tends to become more elastic. The ability of the chains to slip past each other is enhanced and creep phenomena occur more easily at lower stress levels. The elevated temperatures will, at specific critical temperatures for each polymer, cause the crystalline regions to break up or "melt." This temperature is referred to as the *crystalline transition* or *crystalline melting temperature.*

There is a sharp change in properties at the transition. The density undergoes a stepwise increase and the elastic properties change sharply. The polymer becomes an amorphous material and is either a readily deformed solid or, in the case of the materials such as polyesters with strong interchain attraction, a soft rubbery solid. In most cases of good high strength crystalline polymers, the crystalline melting point represents the maximum use temperature to sustain any load.

There is another well defined transition that takes place in the glassy polymers such as polystyrene and polyvinyl chloride. This is referred to as the *glass transition temperature*. The changes that take place are of the same type as in the crystalline melting, but are not as sharply marked or as easily defined as to temperature. There is a fairly sharp reduction in density and a less sharp reduction in the resistance to stress. The most pronounced change is in the resistance to high rate loading, i.e., shock and impact. Above the glass transition temperature the polymers are much tougher although less hard and stiff. There is less resistance to elastic deformation and generally increased retardation in the deformation (greater viscoelasticity) with, in general, the same or even better creep resistance. Some materials exhibit increased creep and the creep levels should be defined in terms of strain levels rather than stress levels to compensate for the reduced elastic constants to determine the true effect. Cooling will have the converse effect that heating has on the properties of materials. By convention the terminology is different despite the fact that the phenomena are the same. When polyethylene is cooled to $-80°$ C it goes through the brittle transition where the material changes from a tough, primarily amorphous material to a rigid brittle material with low impact strength. In general, at lower temperature polymers are more rigid and have higher moduli of elasticity. Because of the reduced chain mobility at lower temperatures, the phenomena of crystallization is usually not seen below room temperature although it has been observed in plasticized vinyl systems where the presence of the plasticizer permits enough mobility of the polymer. In this instance the effect is not reversible upon heating and there is structural damage to the material.

One effect that sustained exposure to elevated temperature has is to stabilize the structure of the material. The processing of polymer materials frequently leaves residual stresses of various types in the material. In addition, the cooling rate on a polymer with a crystalline structure may be such that the equilibrium is not reached during processing. Sustained exposure to heat will result in relaxation of the residual frozen-in stresses and/or the recrystallization of the material. Both of these effects involve movement of the material on a molecular or micro scale with a consequent distortion of the overall shape of the object. In the case of a quenched polyethylene film which has a high degree of clarity, the effect of the recrystallization would be to increase the haze level so that the material becomes opalescent or even opaque.

Probably one of the most important effects of sustained heating on carbon-based polymers is the chemical changes produced in the polymer. Any tendency for further reaction of the polymer chain would be accelerated by the elevated temperature. In some cases the thermal excitation may cause some bonds to break and others to break and reform to produce a thermally crosslinked structure. Since most materials are used in air, the oxidation of the polymer will cause degradation in the physical properties of the material. It would be difficult in limited space to describe all of the effects that heat has on polymer materials. As the design problems are covered, several other effects come up and they will be treated by analyzing the manner in which the basic polymer structure is affected by the thermal excitation.

One of the earliest applications for plastics materials was in electrical insulation. As a result, the electrical properties of plastics have been studied in detail. From our discussion of the structure of the polymer materials we can infer how the properties relate to specific parts of the structure. When a strong D.C. field is applied to several plastics materials, the effect will vary with the structure of the material as can be seen from Fig. 1-19. In the case of a simple material such as polyethylene, there is a mild induced dipole effect and virtually no charge carrier migration. As a result, we can predict that polyethylene is a good electrical insulator with a low dielectric constant and good to excellent resistance to dielectric breakdown. In the case of polyvinyl chloride, there is a strong internal dipole because of the pendant chlorine on the chain so that the applied field causes extensive distortion of the structure. As a consequence, we would expect polyvinyl chloride to have a high dielectric constant. The chlorine attachment to the chain is somewhat labile so that we can expect polyvinyl chloride to be only a fair to poor insulator and readily subject to dielectric breakdown. This is especially true of the unstabilized materials. A polyester material such as is used to make insulating films is a polar material but it has high resistance to loss of the polar groups. These materials have a high dielectric constant but are good insulators and have excellent resistance to dielectric breakdown.

The effect of alternating current fields as shown in Fig. 1-20 is much the same. Polyethylene, with its low polar structure, is affected only slightly by the electric field. The induced dipole is displaced by the periodically alternating polarity of the A.C. field with a minimum of interaction between the molecular chains. Consequently, the dielectric

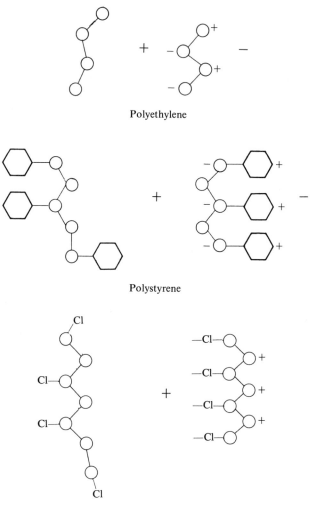

Polyethylene

Polystyrene

Polyvinyl Chloride

Fig. 1-19. Effect of D.C. electrical fields.

loss over a wide range of frequencies is low. Polyvinyl chloride with the highly polar chlorine attached to the chain is strongly affected by the alternating field. The interchain reaction is also quite strong and, consequently, the dielectric loss is high at low to high frequencies. The frictional interaction of the chains appears as heat and is the basis for the dielectric heating process for plastics. In the case of vinyl materials,

Fig. 1-20. Effect of A.C. electrical field on PVC.

it is used to seal the layers of vinyl plastic together to make a variety of products.

The alternating current field also causes additional current flow due to moving ions created by the frictional heating. This leads to poorer resistance to dielectric breakdown under A.C. fields. The effect of the energy absorbtion from the A.C. fields is on both the material and the electrical circuits. In the case of signal carrying circuits, the losses to the insulation can cause destructive loss of signal as well as increased noise due to the effect of heat on the materials.

A tightly structured polymer such as a crystalline polyester or a highly crosslinked polymer where the polar elements can be directly affected by the field by rotation but cannot move far from the equilibrium position does absorb energy from the field and heat to some extent. The restricted motion reduces the effect and these materials show much lower dielectric losses and generally are quite resistant to dielectric breakdown in the A.C. fields. The polyester terephthalate polymer films (some of the most resistant to dielectric breakdown of the commonly used plastics insulating materials), are used in a highly oriented and crystalline form which has the tightest structure and restricts molecular freedom to the maximum extent.

Other electrical effects related to the interaction of electric fields on polymer materials will be covered later in the book as part of the specific design problems. These examples were selected to show the direct action of electric fields on the structure in order to understand how molecular structure is the starting point for the determination of resulting effects.

Electromagnetic radiation covers the spectrum from very long waves of radio energy to the intense energetic gamma rays of very short wavelength. The range can cause a variety of effects on polymer structures, but the examination will be limited at this point to light in the visible, near ultraviolet, and infrared. The most apparent effect of interest in many plastics, particularly the amorphous and glassy materials, is transparency, and plastics are used as transparent structures in many applications such as are shown in Fig. 1-21. Transparency means that the structure will propogate light by the resonant effect of parts of the structure, and the molecular bond lengths and configurations of many plastics are in the proper range for this. Surprisingly, not only are many plastics transparent but they are colorless in that they do not selectively absorb certain wavelengths of light in the visible spectrum.

Color occurs when a specific structure is a selective resonator at a particular wavelength of light or a band of wavelengths. A good way to examine this effect is to take a polymer that is colorless and determine how it can be colored. Polyvinyl chloride is a transparent colorless material. If it is subjected to heat (or alternatively to actinic

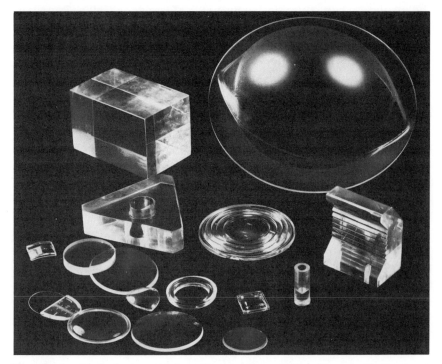

Fig. 1-21. Acrylic plastics are molded, machined and thermoformed. (*Courtesy Glasflex Corp.*)

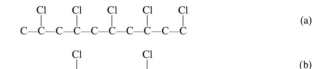

Fig. 1-22. Effect of U.V. exposure on polyvinyl chloride.

light) the structure is changed as shown in Fig. 1-22 and, by the loss of HCl groups, it becomes the unsaturated polymer polyvinylene which had conjugated double bonds. This structure is a resonant absorber of light in the blue range and the material becomes progressively more yellowish and reddish appearing as the reaction proceeds. Polystyrene, which is essentially clear and colorless, does have a slight yellow tint because the styrene structure tends to absorb in the blue and there is a tendency for altered ring structures to form. These act as blue light absorbers.

The velocity of light through any solid is less than that of air and this is the basis of the use of transparent solids as optical elements. It is also true that the velocity of light through a transparent solid will be different for plane polarized light depending upon the orientation of the light with respect to certain internal structures in the material. In the case of plastics, there are two internal structural characteristics that are interesting. One is orientation of the polymer. The second is the effect of strain. Lining up the molecules of a polymer produces a birefringent material which transmits light faster parallel to the orientation than perpendicular to the orientation. A strained amorphous or glassy polymer also exhibits a different velocity parallel to the strain direction. This latter effect is the basis of the rather spectacular frontispiece in this book and illustrates a useful tool in stress and strain analysis using plastics. The optical effect is mathematically related to the degree of strain and a stressed model can be used as an experimental analysis tool in two and three dimensional stress analysis.

The effect of the near ultraviolet light on polymer structure is a chemical one. Many of the structural bonds in plastics materials have an activation energy which corresponds to wavelengths in this range. When these bonds are activated by the light photons, the bonds may be broken, broken and reformed, or new bonds formed. In each case, the structure is changed by the formation of a different polymer arrangement. In some cases the action of the light will cause the release of gaseous products, such as HCl in the case of polyvinyl chloride polymer, or the change in some of the physical properties of the material such as impact strength, tensile strength, or some of the electrical properties. The chemical resistance of the polymer is also altered and this effect has been used to make photo-resist materials.

This chapter has covered the molecular structure of the high polymers which are the materials from which plastics are made. They are basically simple chemical entities which are connected chemically to form long chain molecules and in the case of thermosets, crosslinked from chain to chain. The structure of the polymeric materials has been compared to other types of materials such as metals, ceramics, amorphous materials such as glass, and nonmetallic crystals.

The relationship between the structure of the materials on a molecular level and the gross physical/chemical properties presented a first principles basis for the chapters which follow. These will describe how a polymer or plastics material is designed into useful objects able to withstand the stress and strain necessary in operation and to resist the use environments of the object. The effects of the processing on performance and the limitations of the forming operations will show the designer what can be made effectively from plastics.

Chapter 2

Stress Strain Behavior of Plastics Materials

In order to design a product from any material, it is necessary to have certain basic information both as to the requirement of the part and the response of the material to the conditions of use. The design action required is to select a material and determine the geometry needed to function.

With regard to the stress response of the material, that is, the changes caused by the imposition of loads, several generations of design engineers have been guided by the use of a set of relationships based on Hookes Law, which states that for an elastic material, the *strain* (*deformation*) is proportional to the *stress* (*the force intensity*). Timoshenko and others have developed a literature on the strength of materials based on elastic behavior which has been the basis for the successful design of untold numbers of parts and structures made of metals and other materials that exhibit a good approximation to simple elastic behavior over a wide range of loads and temperatures. In cases where the stress levels are high compared with the strength of the material or repeated loading occurs in use, more sophisticated analysis of the problems was done, and the problems of creep and fatigue were analyzed to deal with this type of application.

Unfortunately, Hookes Law does not accurately enough reflect the stress-strain behavior of plastics parts and is a poor guide to good successful design. Assuming that plastics obey Hookean based deformation relationships is a practical guarantee of failure of the part. What will be developed in this chapter is a similar type of basic relationship that describes the behavior of plastics when subjected to load that can be used to modify the deformation equations and predict the performance of a plastics part. Unlike the materials that have been used which exhibit essentially elastic behavior, plastics require that even the simplest analysis take into account the effects of creep and nonlinear stress strain relationships. Time is introduced as an important variable and, because polymers are strongly influenced in their physical properties by temperature, this too is discussed.

As contrasted to the elastic behavior of metals such as steel and aluminum, plastics behavior is generally described as *viscoelastic*. As the term implies, there is something in the stress response which causes flow and part of the response to the stress is a viscous drag. From our prior discussion of the structure of polymer materials we can visualize where the viscous effect occurs, by polymer chain segments sliding past each other, and where the elastic effect occurs, by the bond bending and direct electrostatic separation of polar portions of the polymer structure. A close look also indicates that even the elastic response which requires the movement of chain segments would be retarded by a drag or viscous effect. After the application of an initial load, it is also apparent on further consideration that as the structure moves in response to the applied load the relationship between the load taken to cause flow and the load taken by elastic deformation will change and that the relationship between the load (stress) and the deformation (strain) will change with time.

In order to analyze these effects, models which exhibit the same type of response to applied forces as the plastics are used which are capable of mathematical formalization. By using appropriate analogy models, the mathematics will accurately reflect the behavior of the real materials.

The elements that are used in such an analysis are a spring, which represents elastic response since the deflection is proportional to the applied force, and the dashpot which is an enclosed cylinder and piston combination which allows the fluid filling the cylinder to move from in front of the piston to behind the piston through a controlled orifice. The behavior of the dashpot is controlled by Newton's Law of Fluid Flow for a perfect liquid which states that the resistance to flow is proportional to the rate of flow.

The retarded elastic response which occurs in plastics materials is best represented as a spring and dashpot acting in parallel. The creep or cold flow which occurs in plastics is represented by a dashpot. Accordingly, the combination which best represents the plastics structure would be a spring and dashpot parallel combination in series with a dashpot. The basic elements and the combinations are shown in Fig. 2-1.

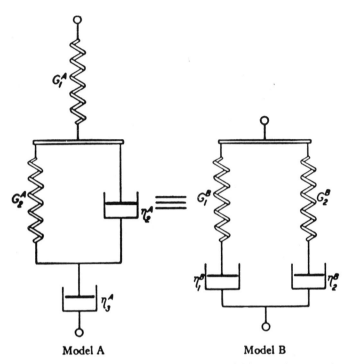

Model A   Model B

Fig. 2-1. Two equivalent models A and B, for material exhibiting elasticity, and flow. (From *Mechanical Properties of High Polymers*, Turner Alfrey, Jr., Interscience, 1948)

The application of this model should be made with the realization of some of the principles implied by the model. These are the concept of the diffusion process which describes the manner in which the chain segments move past each other to give rise to the viscous retardation effects; the assumption of the superposition of the drag and elastic effects (Boltzmann's principle of superposition); and the assumption that the effects are linear and the constants are not affected by the mechanism of stressing the material. Since frictional effects are present, the material tends to heat up during the stressing action and, given the sensitivity of physical properties of plastics to temperature, this is an assumption that must be carefully examined for each case. A method of taking into account the thermal effects is discussed later.

One of the results of the viscoelastic response of polymers is to vary the relationship between stress and strain depending on the rate of stress application. The standard test for many materials to determine structural properties is by analysis of the stress-strain curve which is continued to material failure. The slope of the curve is the elastic constant called the *Young's modulus*, the stress at which the slope of the curve deviates from a straight line is referred to as the *yield strength*, and the stress at which the material fails by separation is called the *ultimate strength*. In the case of viscoelastic behavior, the shape of the curve will depend on the rate of loading or on the rate of straining, depending on the way in which the test is performed. The modulus can vary over a range of three and four to one within the usual testing range and the material can exhibit ductile yielding at the lower straining rates and brittle behavior at the higher straining rates. The value of the yield strength and of the ultimate strength can frequently vary by a three to one ratio.

It is apparent that when tensile tests are done on plastics the loading

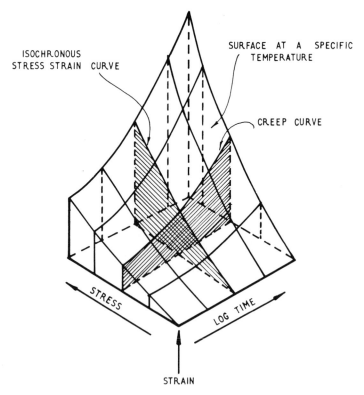

Fig. 2-2. A three-dimensional plot of stress-strain and time. (From *Structural Design with Plastics*, B. S. Benjamin, Van Nostrand Reinhold, 1964)

rates must be specified to make the data have any meaning at all. It also becomes clear that such conventional data are essentially useless for the design of plastics parts unless the in-use loading rates happen to be the same as those of the test. In order to be useful, the tensile tests would have to be run over a wide range of rates and the form in which the data would be presented is a three-dimensional plot of stress-strain and time (Fig. 2-2).

Usually any information on plastics properties is presented in two different forms. One is the stress-strain curve which has been run at loading rates of 25–50 centimeters per minute and is assumed to be the short time behavior of the material. The other set of data is a series of so-called creep curves which present the strain performance of the material as a function of time at several constant stress levels (Fig. 2-3). Neither of these curves is an exact representation of the performance of the material under stress but they represent projections of the

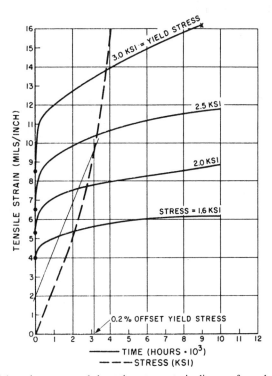

Fig. 2-3. Uniaxial tension creep and short-time stress-strain diagram for polymethylmethacrylate at 73.4 ± 2°F. (From *Engineering Design for Plastics*, Eric Baer, Reinhold, 1964)

stress-strain time curves on the stress-strain plane, and the strain-time plane of Fig. 2-2. The third projection on the strain time plane also represents a situation which is encountered in stress relaxation. An element is strained to conform to a shape and the initial stress decays with time due to relaxation of the material.

Figure 2-4 shows a real test situation in a viscoelastic material. It is a trace of the stress-strain time surface of Fig. 2-2 which shows how the stress application produces a time dependent strain. Curve 1 shown in Fig. 2-4 is one in the range of the standard tensile test and it illustrates the errors that result from assuming the short time test is valid. Curve 2 is the curve that would result from the application of a load in a relatively short time and the subsequent stress-strain history.

We will analyze curve 2 since it represents one of the most frequent load conditions for structural members. The curve shown in Fig. 2-5 is the projection of curve 2 on the strain time plane. The strain at time close to time zero is shown as the intercept on the strain axis. Actually, if the time were magnified in this region, what would be shown is a

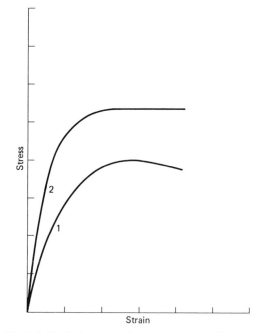

Fig. 2-4. Typical stress-strain curve at two loading rates.

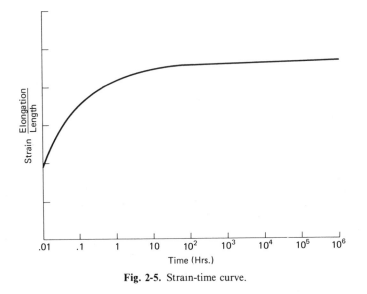

Fig. 2-5. Strain-time curve.

sharply rising strain corresponding to the time of application of the initial stress. As the time increases the strain increases, first, in a highly nonlinear manner (retarded elasticity) then in a monotonically increasing manner (creep). The very initial portion of the curve would correspond to a spring reaction before any of the damped elastic response or creep could take place. The curved portion represents the viscoelastic response timed by the dashpot in parallel with the spring and the rising curve represents the creep resulting from the series dashpot.

Real materials have several mechanisms of elastic response and different chain segments involved in the viscous drag effects related to the different elastic responses. In addition, the creep effects can also have several different sections of the polymer chains flowing in different manners so that there are, in effect, a group of interacting and interconnected spring dashpot sets that are involved in the straining of real materials. There are several ways in which these effects can be analyzed to take account of the various retardation constants and spring constants. As we will see later, this will result in an equation of state that will predict the effective modulus of elasticity of a material that takes into account the change in the strain level with time.

The applied stress field causes a movement of the polymer chain segments past each other as the structure adjusts to the applied load. Turner Alfrey aptly describes this as a stress biased diffusion effect. Since the movement of the molecular segments is a diffusion process, it will generally obey the laws relating to diffusion processes. It will follow an Arrhenius relationship of temperature dependence. It will be directly related to the openness of the polymer structure. It will also depend upon the polar attraction between the diffusing material and the diffusing medium, namely, the shifting chain segments and the remainder of the polymer structure.

The spring constants of the polymer materials will be temperature dependent as well as the retardation constants. Molecular excitation by heating will produce easier bending of the bonds in the chains. In addition, the expansion of the structure will increase the average spacing between the atoms of the structure which will weaken the restoring forces produced by polar and Van der Waals attraction effects.

There are sharp changes in these effects at certain critical temperatures such as the glass transition temperature and the crystalline melting temperature. It has been found that by using these reference temperatures, it is possible to predict the effects of the temperature on the elastic and viscoelastic behavior of the plastics materials.

A typical strain-time diagram at constant stress is shown in Fig. 2-6a. As we indicated before, the strain at any time is the sum of three different mechanisms which can be added by the use of the Boltzmann principle to give the resultant strain.

The first of these is the instantaneous (or short time) strain resulting from the initial imposition of the stress. This strain is shown in Fig. 2-6b. The second strain is produced by the viscoelastic effect or retarded elastic effect which is shown in Fig. 2-6c. The third effect is the creep which is shown in Fig. 2-6d.

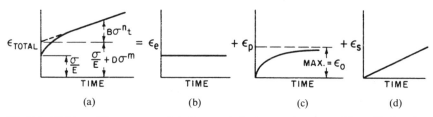

Fig. 2-6. Division of the creep curve into three parts for improved accuracy. (From *Engineering Design for Plastics*, Eric Baer, Reinhold, 1964)

The instantaneous strain shows as a constant level strain which does not vary with time. The viscoelastic stress is shown as a nonlinear increase in strain that levels off after a period of 100 to 10,000 seconds. The creep is a monotonically increasing strain that increases with time. The total strain-time curve is the superposition of the three effects using the Boltzmann superposition theory (Fig. 2-6a). Figure 2-7 shows how the part recovers from the load when the stress is removed. The permanent deformation remains, the instantaneous strain is removed immediately, and the part recovers asymptotically from the viscoelastic strain. The relaxation process is again a diffusion controlled effect and, in general, is the converse effect to the stressing process. It differs in that the biasing effect of the stress field is no longer acting. In general, for calculation purposes, it is assumed to follow the inverse curve to the creep curve.

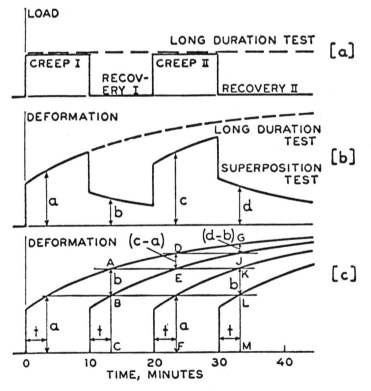

Fig. 2-7. Illustration of the Boltzmann superposition principle (after Leaderman). (From *Mechanical Behavior of High Polymers*, Turner Alfrey, Jr., Interscience, 1948)

It was previously mentioned that the constants of the materials were temperature dependent. It has been shown that the form of the creep curves obtained at elevated temperatures match those taken for longer periods of time at lower temperature. In other words, there is a time-temperature relationship such that creep tests taken for relatively short periods of time at one temperature can substitute for much longer term tests at lower temperatures. This concept is the WLF relationship developed by William, Landel, and Ferry. It is based on a system of reduced variables. The general input from the relationship is that if the creep curves are plotted on a logarithmic scale, the effect of the temperature will be to shift the curve by an amount equal to

$$\frac{\text{The density at a reference temperature times the temperature}}{\text{The density at the required temperature times the required temperature}}$$

Mathematically

$$A_T = D \cdot T / D_0 \cdot T_0$$

where

A_T = shift coefficient
D_0 = reference temperature density
T_0 = reference temperature
D = density at temperature of interest
T = temperature of interest

The literature indicates that for glassy amorphous polymers the best choice of a reference temperature is the glass transition temperature and the density at this temperature. If this temperature is used between the glass transition and the melting point of a polymer it follows that the following equation describes the relationship between the shift coefficient and temperature:

$$\log A_T = \frac{-C_1(T - T_G)}{C_2 + (T - T_G)}$$

where C_1 and C_2 are approximately universal constants with values of 17.4 and 51.6, respectively, and T_G is the glass transition temperature. It is noteworthy that the relationship holds well with nonpolymer materials and crystalline polymers which also have a glassy region for a range of 100°C above the glass transition temperature.

We will summarize in mathematical form the above discussion, but first, the salient points will be made. Plastics do not respond elastically to stress but in a more complex manner which includes elastic, retarded elastic or viscoelastic, and viscous strain (or creep). These effects are essentially diffusion processes when it comes to the viscous and viscoelastic behavior and the effects are highly temperature and density dependent. By the use of spring and dashpot models we can show the behavior of these materials and, using the Boltzmann principle of superposition, we can add the several effects and produce a time-dependent strain curve at constant stress. A family of these curves will describe the performance of a material if used in conjunction with a short time stress-strain curve. We find that time and temperature have an interdependence so that a short time test at elevated temperature can be transposed to a long time test result at lower temperature using the WLF relationship. In addition, we find that the relaxation curve can be approximated by the inverse of the creep curve and that the curves for real materials can be generated by summations of the effects of large collections of springs and dashpots of different constants using the superposition principle of Boltzmann.

This collection of principles enables us to use short time tests on materials to construct the necessary creep data for a variety of in-use conditions and times and stress cycles to predict the strain performance of plastic materials. It is hardly as simple as Hookes law, but it is a manageable collection of relationships that show how the geometry of a part is altered by stress and is used to determine the necessary geometry to provide a functionally useful part given the required stress it must withstand.

In the relationships which follow, which are given without proof except by reference, the stress fields that are used in the analysis are shear stresses rather than tensile or compressive stresses. These are used to simplify the mathematics of the derivations. The tensile stress is rather simply related to the shear stress as can be seen in Fig. 2-8. A differentially small element of a material is shown with a tensile stress applied across the corners of the element. This force I is opposed by the forces generated by the material forces F_a, F_b, F_c, and F_d. The element deforms as a result of the force imposed by the angle a. $F_a = F_b = F_c = F_d = F_e$.

The shear modulus G is related to the tensile modulus E by the relationship

$$G = \frac{E \cos \dfrac{90 - a}{2} - \cos 45°}{\sin a \cos \dfrac{90 + a}{2}}$$

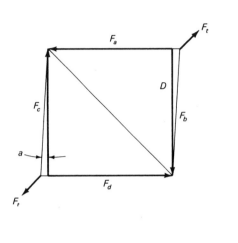

$$F_a \cos (90 - a) + F_b \cos (90 - a) = E_t$$

$$\text{Shear strain} = D \sin a$$

$$\text{Tensile strain} = 2D \cos (90 - a)$$

$$\frac{\text{Shear stress}}{\text{Shear strain}} = \text{Shear modulus} = G$$

$$\frac{\text{Tensile stress}}{\text{Tensile strain}} = \text{Tensile modulus} = E$$

$$G = \frac{F_a}{D \sin a}$$

$$E = \frac{2F_a \sin 45°}{2D \cos (90° - a)}$$

$$\frac{G}{E} = \frac{\cos (45° - a)}{\sin (45° - a)} E$$

Fig. 2-8. Relationship between tensile stress and shear stress.

See Fig. 2-4 for the geometrical relationships in a two-dimensional analog of the three-dimensional stress-strain condition. So far the discussion of stress-strain relationships has been on static stress conditions. Cyclical stresses such as are applied in the operation of machine parts, in vibrating structures, and in the rapid flexing of structures that occurs in vehicles produce a different set of stress response problems.

One of the significant situations is that when the stresses are applied for periods of a small fraction of a second to, at most, a second there is little creep effect. The response of plastic materials is elastic and viscoelastic. In this context, it is interesting to look at the stress-strain curve in the loading and unloading. Figure 2-9 shows a curve for a typical viscoelastic material undergoing short-term loading and unloading at several high loading rates. The recovery curve which generally returns to zero strain indicating no creep does not follow the loading curve. The area between the two curves represents mechanical energy which has been absorbed by the material. While some of this energy may have been used to produce molecular level structural changes such as bond rupture or changes in the crystallization of the crystalline portions of the polymer, the bulk of the energy is converted to heat by the friction produced between the chain segments moved by the stress application. Since the physical properties of

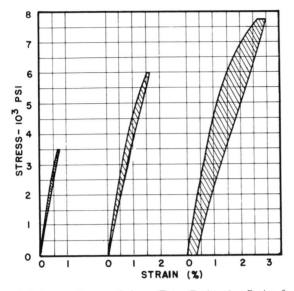

Fig. 2-9. Hysteresis tests on polyoxymethylene. (From *Engineering Design for Plastics*, Eric Baer, Reinhold, 1964)

polymer materials are sensitive to temperature it follows that this heating effect can profoundly affect the performance of the material. The availability of stress-strain data under varying load is important for the design of parts subjected to this type of stress.

It is interesting to note that the loading rate can be selectively more effective at certain rates for specific materials (see Fig. 2-10). If the rate is in the range of a time constant for one of the retardation modes in the structure of the polymer the energy absorbtion rate is substantially higher than at other loading rates (Fig. 2-11).

The stress level at which plastics materials fail under rapid repeated stress is always lower than the level at which they fail under short time high level stress and usually well below the level at which long time stress is applied. In addition to the heating effect previously mentioned, there are two other effects of cyclical loading that lead to failure. One of these is the rupture of bonds in the structure due to localized high level stress. The second is other structural changes such as recrystallization processes due to local movement of the polymer chain segments. Whether the material fails due to the heating effect or to the type of work-hardening-effects, such as bond breaking or recrystallization, depends on the heat generating characteristics of the

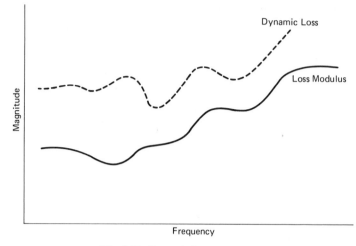

Fig. 2-10. Dynamic frequency spec.

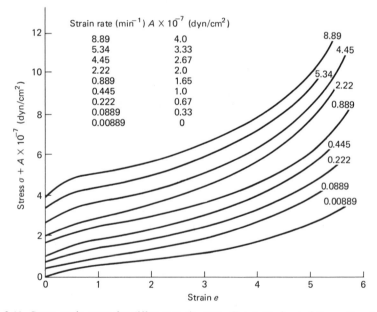

Fig. 2-11. Stress-strain curve for different strain rates. (From *Mechanical Properties of Solid Polymers*, I. M. Ward, Wiley-Interscience, 1971)

material, the stress levels that are attained, and the ability of the part to dissipate the heat generated by the cyclical loading. Given a part with geometry that permits efficient heat transfer (such as a thin strip) and a material having low hysteresis effects (such as a crystalline polyester polymer), the probable mode of failure will be a true fatigue effect.

In most instances of plastics failure under cyclical loading, the hysteresis effects are the ones that cause failure. The failure in this case occurs in a fairly short time. The cyclical loading causes the material to heat up, reducing the modulus of elasticity and increasing the hysteresis loop in the stress-strain curve which in turn increases the rate of heating. This results in yet higher temperatures, etc., and catastrophic failure by severe softening of the material occurs. Only by improving the rate of heat transfer to the surroundings so that a stable temperature is reached below the softening point of the material is it possible to avoid this type of failure. The fatigue failure of plastics materials under cyclical loading follows the same type of stress/number of cycles or SN curve as do other materials. A typical set of curves of this type is shown in Fig. 2-12.

To our principles for static or long-term loading we have added those which apply to short-time cyclical loading such as is encountered

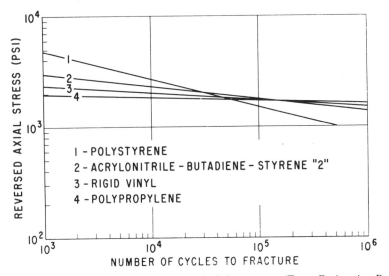

Fig. 2-12. Typical examples of different slopes in fatigue curves. (From *Engineering Design for Plastics*, Eric Baer, Reinhold, 1964)

in machine parts and in vibration environments. The most significant of these is the effects caused by the energy absorbtion in the material as a result of hysteresis. This energy is dissipated as heat primarily, which causes the material to rise in temperature and, if not balanced by heat dissipation to the surroundings, leads to catastrophic failure by softening of the material. The other means of absorbing the hysteresis energy is by bond fracture and by recrystallization of the plastics structure which is analogous to the work hardening phenomena that occur with other materials that lead to typical fatigue failure effects.

Chapter 3

Effects of Fillers
on Properties and
Performance

A filler is defined as any material added to a polymer that is significantly different structurally or chemically from the basic polymer. By convention this exempts such materials as plasticizers or compatible resins which in effect become part of the continuous plastic phase. Even the addition of a rubbery second phase such as exists in high impact polystyrene is excluded although it can be distinguished from the polystyrene phase microscopically because the material acts in most respects as a homogeneous phase.

The fillers used cover a wide variety of materials. Some of the typical ones are listed in Table 3-1. They can be grouped in several ways; in terms of their structure, the way they interact with the polymer matrix, or in terms of their composition. Structurally the fillers may be aggregates with essentially round or polyhedral shapes such as clay or chalk, plates or flakes such as mica, or lamellar glass, fibers such as fiber glass, asbestos, or synthetic fibers, or cellular material such as vermiculite, foamed glass, or hollow glass beads.

The second way that the fillers can be characterized is by their interaction with the polymer matrix. The fillers can be adherent to the polymer matrix whether inherently or by special surface treatment. The fillers may absorb the polymer phase because of high surface area and inherent wettability. They can react chemically with the polymer material to form a chemical bond, again, both by inherent structure or by suitable surface treatment. The filler can be catalytic to the structure and, in addition to acting as a filler can cause crosslinking of the structure, notably in the case of carbon black. Finally, the filler can be inert and nonadherent to the polymer and remain as just a void filler.

The chemical nature of fillers can cover a wide range, some of which have been indicated. Carbon black is an inorganic residue

Table 3-1. Some Fillers and Reinforcements and Their Contribution to Plastics

Filler or reinforcement	Chemical resistance	Heat resistance	Electrical insulation	Impact strength	Tensile strength	Dimensional stability	Stiffness	Hardness	Lubricity	Electrical conductivity	Thermal conductivity	Moisture resistance	Processability	Recommended for use in[a]
Alumina tabular	●	●				●								S/P
Alumina trihydrate, fine particle			●			●						●	●	P
Aluminum powder										●	●			S
Asbestos	●	●	●	●		●	●	●						S/P
Bronze							●	●		●	●			S
Calcium carbonate[b]	●					●	●	●					●	S/P
Calcium metasilicate	●	●				●	●	●				●		S
Calcium silicate	●					●	●	●						S
Carbon black[c]	●					●	●			●	●		●	S/P
Carbon fiber										●	●			S
Cellulose				●	●	●	●	●						S/P
Alpha cellulose		●		●	●									S
Coal, powdered	●												●	S
Cotton (macerated/chopped fibers)		●		●	●	●	●	●						S
Fibrous glass	●	●	●	●	●	●	●	●					●	S/P
Fir bark													●	S
Graphite	●				●	●	●	●	●	●	●			S/P
Jute			●			●								S
Kaolin	●	●				●	●	●	●			●	●	S/P
Kaolin (calcined)	●	●	●			●	●	●				●	●	S/P
Mica	●	●	●			●	●	●	●			●		S/P
Molybdenum disulphide							●	●	●			●	●	P
Nylon (macerated/chopped fibers)	●	●	●	●	●	●	●	●	●				●	S/P
Orlon	●	●	●	●	●	●	●	●				●	●	S/P
Rayon		●	●	●	●	●	●							S
Silica, amorphous		●										●	●	S/P
Sisal fibers	●			●	●	●	●	●					●	S/P
TFE—fluorocarbon							●	●	●	●				S/P
Talc	●	●	●			●	●	●	●			●	●	S/P
Wood flour		●				●	●							S

*The chart does not show differences in degrees of improvement; calcined kaolin, for example, generally gives much higher electrical resistance than kaolin. Similarly, differences in characteristics of products under one heading, such as talc (which varies greatly from one grade to another and from one type to another) also are not distinguished. a—Symbols: P—in thermoplastics only; S—in thermosets only; S/P—in both thermoplastics and thermosets. b—In thermosets, calcium carbonate's prime function is to improve molded appearance. c—Prime functions are imparting of U-V resistance and coloring; also is used in cross-linked thermoplastics.

*Courtesy "Modern Plastics Encyclopedia."

of the pyrolysis of organic materials. Clays, silica, and chalk are fine polycrystalline mineral materials. Mica and asbestos are structured mineral-type materials which are again basically crystalline in nature. Synthetic mica is a 2000°F mica. Fiberglass and fused silica fibers are amorphous glassy materials drawn into fibers. In addition, there are graphite fibers which are highly structured crystalline carbon, synthetic fibers such as polyesters, polyamides, polyimides, and other fiber types. We also have the cellular materials which can be made from glasses such as glass beads, from polymer materials such as phenolic microspheres, and finally, foamed minerals such as vermiculite, foamed glass, and foamed plastics such as urethane and phenolic foam.

In most of the discussion that follows, the principal characteristics of the improvements to be expected from the incorporation of fillers result from the fact that they react with and are adherent to the structure. The nonadherent situation has just recently assumed an important role in the manufacture of films to be used in printing and other paperlike applications. The preparation of these films requires a degree of microporosity so that they can absorb conventional inks and pass through water vapor and other gases. To produce this effect, fillers are added to the plastics materials which are nonadherent to the polymer structure. For example, ground untreated glass is added to polyethylene. This compound is converted to a film and subsequently oriented during the processing. Because of the stresses around each of the nonadhering filler particles during the orientation process, the polymer separates from the filler to form a micro-void, and the entire structure is microporous. The untreated filler is readily wetted by conventional water based and similar inks, and the combination produces a good paper substitute with excellent stability and permanence characteristics as compared to cellulose based papers. There are undoubtedly other applications where this phenomenon can be employed effectively.

In general, the addition of nonadherent filler is an undesirable method of filling plastics materials. The fillers add bulk and improve some properties such as the compressive strength and modulus, and they generally improve the heat resistance and reduce the coefficient of thermal expansion. Because of the nonadherence, however, they make the materials brittle and weak in tension and bending. This results from the fact that each particle acts as a stress riser, increasing the stress levels around each particle so that premature failure will occur (Fig. 3-1). The properties of a material with a nonadhering filler are equal to, but not greater than, the properties of the resin phase, and, if the material·is notch-sensitive, as are most polymer materials, the properties are poorer than those of the resin phase. If a material is filled with 50% by volume of a nonadhering filler, the physical properties will reach a maximum of 50% of the value of the unfilled resin.

The case of fillers which adhere to the polymer matrix is important in most applications. This is general whether the adhesion is intrinsic to the materials or enhanced by surface treatment of the fillers or additives to the polymer. When the adhesion between the filler and the resin matrix is adequate, the resin and the filler coact under stress. From the standpoint of visualizing how these effects alter the performance of the composite to make the whole greater than the parts, we will examine the combination of a fibrous filled fiber glass and a suitable resin matrix such as an adhering polyester resin. The case of a simplified configuration of a bundle of parallel fibers before and after the addition of the resin matrix is illustrated. The analysis will be extended to random fiber orientation and then to other filler shapes.

Figure 3-2 shows a bundle of fiber glass filaments and the effect of applying tensile, compressive, bending, and shear stresses to the bundle. In the case of the tensile and compression loads, they are applied axially along the direction of the fibers. It is apparent that the fibers cannot take any compressive load since the tiny fibers have no column stability at all and buckle at low load. In tension, if the load is applied over a length where the fibers are continuous, the material is very strong and the bundle will exhibit almost the

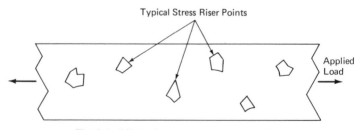

Fig. 3-1. Effect of stress on nonadherent filler.

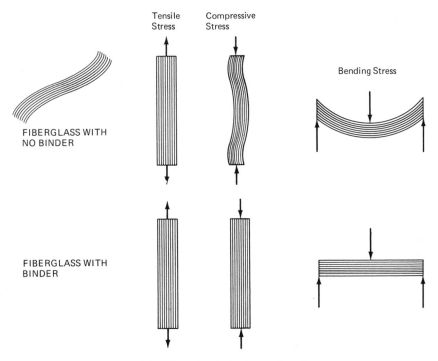

Fig. 3-2. Effect of resin binder on fiber glass.

maximum strength of the filaments. The material has no significant elongation so that it will undergo brittle failure. Transverse loading of the bundle of filaments either by bending or shear is almost impossible since the small diameter filaments acting individually bend at low stress levels and it is difficult to apply a shear load without tightly gripping the bundle of filaments which would tend to crush them and give very low values of shear strength.

When the resin matrix is added to the filaments, the situation is drastically altered. The rod formed when the fiber glass filaments are bonded together is far better then either the resin matrix or the fiber glass filler. Under compressive loading the material is able to take a load equal to the sum of the compressive strength of the glass fiber and the resin matrix. It is not necessary that the filaments be continuous from one end of the rod to the other since the resin phase acts as a stress transfer medium to transfer the stress around discontinuities in the filaments. In addition, the resin phase will yield and give a small but reasonable deformation before fracture.

The same argument may be applied to tensile stresses. The composite structure will have tensile strength and tensile modulus which reflects the values for the glass plus the resin phase. The stress transfer around the discontinuities in the filaments will increase the average tensile strength to the glass phase. In addition, the yield in the resin phase will overcome the tendency to brittleness in the material.

The resin fiber glass composite is now able to take transverse loads as well as tensile and compressive loads. By referring to Fig. 3-2, it is apparent that bending, which is a combination of tensile and compressive stresses on opposite sides of the member being loaded, is easily taken by the resin bonded fiber glass filaments. The strength in bending as well as the modulus is the sum of the values for the two components of the system. In this case, as well as the others for compression and tension, the stress transfer function of the resin phase contributes higher strength by enabling the stress to be transferred around discontinuities in the glass fibers, as well as adding resiliency to the structure to avoid brittleness. It is also evident that the material will take shear stresses equal to the sum of the shear strength of the resin and glass components.

The composite structure has a much lower creep level than the resin matrix material. This results from several effects, the first of which is the rather obvious one of bulk. A substantial portion of the combination is made of a low creep material, glass, and, consequently creep would be reduced. A major part of the reduction in creep, however, comes from other characteristics such as the chain lengths involved in taking the stress from one fiber to the other. Figure 3-3 shows schematically the coupling length of a polymer chain that connects

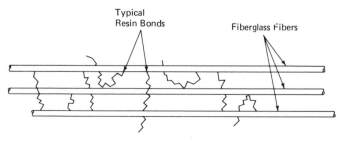

Fig. 3-3. Mechanism of bonding of fiber glass with resin showing short chain length between bond points.

two glass fibers. Under most loading conditions, this chain segment is loaded in shear as the two fibers try to slide past each other. The effect on the chain segment is to stretch it as if it were a crosslinking element in the composite structure. The chain tends to straighten out by bond bending and biased diffusion effects with a minimum of interchain friction effects. In addition, since the chain segments are bonded at the ends to the glass fibers, they do not slide past each other in the creep mode. It is necessary for the bond between the polymer chain segment and the glass fiber to be broken before the creep can take place to any significant extent. It is obvious that the higher the proportion of glass to resin, the shorter the average chain segment to the bond point, and the lower the creep and viscoelastic response. At high glass loadings the material becomes fairly elastic and behaves like other classic engineering materials such as glass and metals.

The change in properties extends beyond the static properties of the material. Under dynamic loading, the material behaves much more elastically and the tendency to destruction by hysteresis induced heating effects is reduced. On the other hand, because the polymer chains are stretched severely in the stress transfer process, there are much higher levels of bond breakage and recrystallization effects with typical fatigue failure performance. In the balance, the net result is a substantial improvement in the performance of the material in dynamic loading.

The thermal properties of the material are substantially altered by the fiber filler. The thermal conductivity of the composite is increased to a level close to that of the glass filler. This improves the performance under dynamic loading by increasing the heat transfer through the material and reducing the tendency to local heating and failure. The thermal expansion of the material is drastically reduced. Fiber-glass resin composites have coefficients of thermal expansion from 10 to 20% that of the resin matrix. In the section on processing we will find that the fiber glass has a substantial effect in improving moldability. In addition, the performance temperature of the composite is raised substantially and the general stability of the combination is much better than that of the resin matrix.

The analysis has been limited to materials having the fibrous glass filler present in long monofilaments parallel to each other. This is the most efficient use of the filler from a structural standpoint provided that the stress direction is parallel to or perpendicular to the fiber direction and depending on whether it is tension or compression for the first and bending or shear for the second. In many applications it would be better if the material were isotropic in one plane or even, for the more general case, isotropic in all directions. This can be done by the use of filaments arranged in staggered angle directions in the sheet of the material, by the use of woven materials, or, for the completely isotropic material, filaments randomly dispersed in the material. It is obvious that the fiber glass is not used as efficiently in these arrangements and the increase in physical strength properties is significantly less in any one direction than it is for the parallel fibers. It is also obvious that the arguments used for the improvement in properties in the unidirectional case apply to the other reinforcements. An analysis of the random fiber condition and for the woven reinforcement is given in Milewski.* These give the quantitative improvement in the physical properties of fiber-glass reinforced resin composites.

Other types of fibers are used to reinforce plastics and a partial list is given in Table 3-2. They range from mica and asbestos, a naturally occurring mineral fiber, to such materials as cotton and linen, to synthetic fibers like polyesters and polyamides, and finally, to unusual reinforcing agents such as graphite fibers, boron metal fibers, and synthetic gem whiskers like sapphire. These other reinforcing fibers range widely in elastic properties and temperature resistance; they impart other properties such as desirable electrical values and resistance to aggressive environments. Graphite is one of the most unusual of the reinforcements in that it imparts excellent high temperature resistance, controllable thermal expansion, electrical conductivity, and a very high elastic modulus. It also may permit controlled thermal expansion. The argument for improvement of properties is similar to that of the fiber glass but the values are different.

The use of platelike or flake reinforcement is common in plastics composites. Mica, both natural and synthetic, is one of the most common platelike reinforcements. Another is flake glass. The effect of this type of reinforcement depends again on the orientation of the

---

*Chapter 12, John V. Milewski in *Handbook of Fiberglass and Advanced Plastics Composites*, edited by George Lubin, Van Nostrand Reinhold, New York, 1969.

Table 3-2. Property Comparisons Unreinforced and Reinforced Materials. (Columns marked U = Unreinforced, R = Reinforced.)

	Polyamide (Nylon) U	R	Polystyrene* U	R	Polycarbonate U	R	Styrene Acrylonitrile† U	R	Polypropylene U	R	Acetal U	R	Linear Polyethylene U	R
Tensile strength (1000 psi)	11.8	30.0	8.5	14.0	9.0	20.0	11.0	18.0	5.0	6.6	10.0	12.5	3.3	11.0
Impact strength, notched (ft-lb/in)														
At 73°F	0.9	3.8	0.3	2.5	2.0§	4.0§	0.45	3.0	1.3–2.1	2.4	60.0	3.0	—	4.5
At −40°F	0.6	4.2	0.2	3.2	1.5§	4.0§	—	4.0	—	2.2	—	3.0	—	5.0
Tensile modulus (10^5 psi)	4.0	—	4.0	12.1	3.2	17.0	5.2	15.0	2.0	4.5	4.0	8.1	1.2	9.0
Shear strength (1000 psi)	9.6	14.0	—	9.0	9.2	12.0	—	12.5	4.6	4.7	9.5	9.1	—	5.5
Flexural strength (1000 psi)	11.5	37.0	11.0	20.0	12.0	26.0	17.0	26.0	6 to 8	7.0	14.0	16.0	—	12.0
Compressive strength (1000 psi)	4.99††	24.0	14.0	17.0	11.0	19	17.0	22.0	8.5	6.0	5.2	13.0	2.7 to 3.6	6.0
Deformation, 4000 psi load (%)	2.5	0.4	1.6	0.6	0.3	0.1	—	0.3	—	6.0	—	1.0	—	0.4‡
Elongation (%)	60.0	2.2	2.0	1.1	60–100	1.7	3.2	1.4	>200	3.6	9–15	1.5	60.0	3.5
Water absorption in 24 hr (%)	1.5	0.6	0.03	0.07	0.3	0.09	0.2	0.15	0.01	0.05	0.20	1.1	0.01	0.04
Hardness, Rockwell	M79	E75–80	M70	E53	M70	E57	M83	E65	R101	M50	M94	M90	R64	R60
Specific gravity	1.14	1.52	1.05	1.28	1.2	1.52	1.07	1.36	0.90	1.05	1.43	1.7	0.96	1.30
Heat distortion temp at 264 psi (F)	150	502	190	220	280	300	200	225	155	280	212	335	126	260
Coefficient of thermal expansion (per °F $\times 10^{-5}$)	5.5	0.9	4.0	2.2	3.9	0.9	4.0	1.9	4.7	2.7	4.5	1.9	9.0	1.7
Dielectric strength, short time (v per mil)	385	480	500	396	400	482	450	515	750	—	500	—	—	—
Volume resistivity (ohm-cm) $\times 10^{15}$	450	2.6	10.0	36.0	20.0	1.4	10^{16}	43.5	17.0	15.0	0.6	38.0	10^{15}	29.0
Dielectric constant at 60 cps	4.1	4.5	2.6	3.1	3.1	3.8	3.0	3.6	2.3	—	—	—	2.3	—
Power factor at 60 cps	0.0140	0.009	0.0030	0.0048	0.0009	0.0030	0.0085	0.005	—	—	—	—	—	—
Approximate cost (¢/cu in)	3.6	8.8	0.5	2.5	4.5	7.9	1.0	3.5	0.8	2.1	3.3	7.8	0.7	3.1

*Medium-flow, general purpose grade. †Heat-resistant grade. §Impact values for polycarbonates are a function of thickness. ‡1000-psi load. ††At 1% deformation.
From Rexall-Fiberfil data.

flakes and on the quality of the resin reinforcement bond. If the bond is strong enough in shear so that the reinforcing material can be stressed to a large fraction of the tensile or compressive strength of the composite it is substantially greater than the parent resin material.

Our analysis again will deal with the simplest arrangement of the reinforcing material to show clearly how the effect is achieved. The arrangement that gives the maximum reinforcement is shown in Fig. 3-4 where the flakes lie parallel to each other in layers. The end view of this arrangement looks similar to the parallel filaments in Fig. 3-2 and it can be seen that again the resin phase holds the reinforcement together with short segments of the polymer chain. The stress pattern is somewhat different than in the case of the filament because of the layer effect on the directionality, and the interplane adhesion is stronger than the filament-to-filament adhesion. In any event, the improvements in compression and tension properties in the direction of the flake planes is obvious, and consequently the bending performance is also improved.

Unlike the filament case there are different properties in the three mutually perpendicular planes that can be used to define stress direction. The properties in the planes perpendicular to the flake plane show maximum strength in bending and shear as well as compression. The material would be weak in tension in these planes. In the plane of the flakes, the material would be strong in compression and tension and weak in bending. The shear strength would be equal to the shear strength of the bonding resin. In the plane perpendicular to the flake direction, the material would be strong in compression and tension, fair in bending, and good in resistance to shear. This is evident from Fig. 3-5 which shows how the filler is aiding in the stress handling situations.

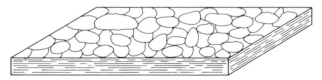

Fig. 3-4. Flake or platelet laminate.

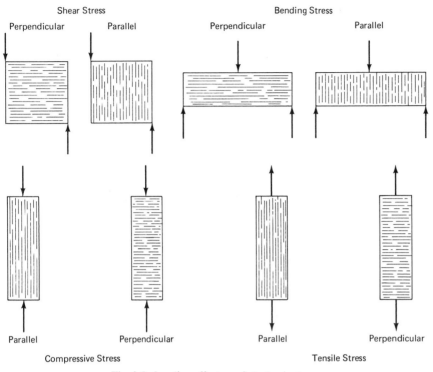

Fig. 3-5. Loading effects on flake laminates.

In actual materials the flakes are rarely laid down in the orderly manner shown here. However, the natural tendency of these materials is to form layers when molded or otherwise converted to a product. The properties found in a finished product are easily deduced from the properties of a parallel layered material as achieved in the various sections of a complex part. An analysis of the flow patterns in the processing will show where the physical properties exist and in what directions.

The nonfibrous polyhedral fillers which have good adhesion to the resin matrix also contribute significant improvements to the overall physical properties of the composites of which they are a part. The effect is not as dramatic as that for the extended surface fillers such as the fibers and flakes, but they do represent an important way of improving the performance of plastics, usually with a reduction in cost.

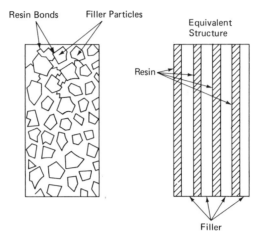

Fig. 3-6. Filled plastics structures.

Figure 3-6 shows a small segment of a structure containing an adherent polyhedral filler material responds to the stresses of tension, compression, bending, and shear. In compression, the material will show a substantial improvement in both modulus and ultimate strength. The material divides the stress between the filler and resin phases so that the combined value of the two appears to be the volume average of the values for the two phases. The interaction is a simplified one in which we have shown parallel columns of resin and filler and resin alone. In a random dispersion, this arrangement would not exist where the stress transfer is by direct compression; however, the stress transfer by shear plus compression shown in Fig. 3-7 would give essentially the same results.

Tensile stresses applied to the filled material would be resisted by columns of material alternatively of resin and filler. The modulus would at first glance be a distance-weighed average of the filler and the resin and this, in fact, is the major effect. However, the resin chain segments between the filler particles are relatively short and at small strain levels they would be stretched to a high degree. As a consequence, the resin phase will act much more elastically and show a higher modulus of elasticity than the bulk material. The tensile modulus would be somewhat higher than the distance-weighed average calculation would indicate.

The tensile strength of the material would be limited by either

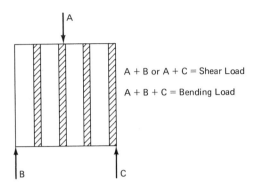

Fig. 3-7. Filled plastics structure under load.

the cohesive strength of the resin phase or the adhesive strength of the resin to the filler. The short segment bonding chain lengths which connect the filler particles would tend to stretch more for a given strain level so that the cohesive strength of the material would be higher than the bulk material. In most cases, the limiting factor in tensile strength would be the adhesive bond between the resin and the particulate filler.

The modulus of elasticity in bending would be an average of the tensile and compressive moduli since the bending involves fiber stress of both types. Since the compressive modulus would tend to be higher from our argument above, it is apparent that the neutral plane in bending would not be in the midplane of the bent member which will affect the deflection of the member. The ultimate strength in bending would also be limited by the tensile strength since it would be lower than the compressive strength for this type of composite material.

Since the performance of particulate filler reinforcements depends to a major extent on the adhesive bond between the filler and the resin phases, it is apparent that different fillers will have substantially different degrees of effect on the properties. This would be in addition to the strength and stiffness characteristics of the filler structure. A filler material that has a high surface area as compared to its volume would, other factors remaining constant, have better adhesion to the resin and give better strength. Colloidal silica and carbon blacks are two typical reinforcing fillers that have exceptional effects in improving the properties of a resin filler composite. Some filler materials are actually porous and the resin phase interpenetrates the filler to give unusually good improvements in properties. Certain forms of aluminum oxides made from solution exhibit this microporosity and for some resin systems are especially effective filler materials.

Many filler materials are treated chemically to improve the adhesion between the filler and resin. The most widely known of these treatments are used on glass fibers and include silane reaction products and chromium complexes. Glass fibers are often flame treated. Several of the treatments are summarized in Table 3-3. Improved versions of these treatments have resulted in major improvements in the performance of fiber-glass reinforced plastics so that a large fraction of the potential strength of these materials is realized in practice.

Calcium carbonate, another widely used filler for both thermoplastic and thermoset resins, is available in a variety of special surface treatments which enhance the wettability by resins to improve the physical properties of the composites to which it is added.

In many cases several different fillers are used in a composite material. The use of a clay with fiber glass is one typical combination. The clay is used to add bulk and improve primarily the compression properties while the fiber glass is used to improve the tensile and compression properties. The relatively inert extender also reduces the coefficient of thermal expansion and improves the thermal conductivity which is advantageous in many cases. Colloidal silica is also frequently added to glass reinforced materials to improve the tensile properties by stiffening the resin phase. Beryllium fillers greatly increase the thermal conductivity.

The most economical filler for a resin is a gas such as nitrogen, or air. Many low density plastics are made by incorporating a gas during the processing as, for example, in thermoplastic and thermoset foams like urethane and styrene foam. The other approach is to use a cellular filler such as vermiculite or phenolic foam or microballoons made from glass or phenolic resins. The use of a low density filler can improve the physical properties or reduce them, depending on the nature of the filler. The use of material such as glass microballoons will improve the rigidity of the composite as well as the strength at the same time that it reduces the density. Other porous fillers such as chopped foam and direct injection of gas will

Table 3-3. Commonly Used Commercial Glass Fiber Finishes.

Finish 111	Part of the starch sizing on the yarn is burned off and the remainder is caramelized. The fabric has a tan color and a residue of sizing. This finish is especialy suited for melamine laminates.
Finish 112	Almost all the organic sizing is burned off the fabric. The fabric is white and virtually pure glass. It is the intermediate finishing stage in the production of fabrics with chemical finishes which provides protection, resin compatibility, and high performance of laminates. Finish 112 is used, as is, in the production of silicone laminates for improved electrical properties.
Finish 112-Neutral PA	This finish is similar to 112 except that after heat cleaning, the fabric is washed in demineralized water to remove residual alkali. Such treatment improves the electrical performance of silicone laminates.
Finish 114 or "Volan"	After heat cleaning (Finish 112), the fabric is saturated with a chrome complex or "Volan" (trademark). Fabric with 114 finish is compatible with polyester resin.
"Volan A" Finish	"Volan A" is a modification of Finish 114 that provides better laminate wet strengths. It is the most widely used fabric finish for laminating with polyester, epoxy, or phenolic resins. "Volan A" finish is obtained by saturating 112 finished fabric with methacrylate chrome chloride, curing, and then washing to remove any remaining soluble salts.
Finish 550	This finish is an improved "Volan" designed to provide laminar wet strength retentions above that of "Volan A." It also seems to give improved dry laminate strengths. It is compatible with polyester and epoxy resins.
A-1100	The fabric is heat cleaned (Finish 112) and saturated with an amine-type vinyl silane. It is used with epoxy, melamine, and phenolic resins and as a finish provides good high temperature properties. A-1100 can be provided as a "hard" finish with a stiff hand or as a "soft" finish which is pliable and drapeable.
"Garan" Finish	"Garan" (trademark) is a vinyl silane applied to 112 finished fabric. It is compatible with polyester resins and can be supplied as a hard or soft finish.
901 (HTS) and 904	These epoxy compatible treatments are applied to S glass filaments at the bushing. They provide superior high strength performance and improved laminate wet strength retention. Finish 901 requires cold storage; finish 904 is not temperature sensitive.
LHTS	LHTS is a series of epoxy compatible treatments applied at the bushing to E Glass or S Glass filaments. They are similar to 901 and 904 in that they provide high strength performance and improved wet strength retention.

*Trademark of Owens-Corning Fiberglas Corp. Patent Number 3,312,569.

generally lower the physical properties, although the strength-weight ratio is always improved, particularly in well designed components that combine solid and cellular materials.

Another class of fillers that is frequently used includes metal fibers and metal wire and wire products. The usual reason for adding these fillers is for the conductivity they add to the material. There has been recent work in which metal fibers have been used in a manner similar to glass fiber but for reasons of cost and weight, the metal fiber does not seem to be a good reinforcing fiber. Ultra high strength single crystal whiskers may be one way in which the fiber metal materials serve to advantage although practical utilization seems to be some time off.

In summary, we have seen that the incorporation of fillers of various types and structures can impart significant improvements to plastics components. They act as stiffening agents and improve the physical strength and resistance to creep and flexural failure. They may serve as low cost extenders. The various forms have differing degrees of efficiency in improving the properties of the composites with filler resin matrix adhesion as the primary requirement for the structural improvement. Although this is the primary usage, some interesting effects can be obtained by non-adhesion making the filled plastics one of the more important and interesting group of plastics materials available.

Chapter 4

Stress Analysis for Plastics

The preceding chapters are intended to provide background for the design of plastic parts. In this chapter we will apply the concepts to the actual design process and show how to make a design that will function. Typical plastics parts are shown in Figs. 4-1, 4-2, and 4-3.

This is best illustrated by a simple example that utilizes the steps required in most design problems and shows how to apply the methods with the available design data to make an efficient design. Figure 4-4 shows a sketch of a plastics link used to support the front part of a hinged shelf. The part is stressed in simple tension and the primary function of the part is to maintain the shelf level as it supports part of the weight of the shelf and its contents. There are several obvious geometrical requirements of the part. The link must be long enough to reach from the upper support point A to the normal open position of the shelf. The thickness of the link should be in a range such that it will not interfere with another shelf hanger at an appropriate distance from the first so that a wall of shelves can be made. The width of the link should be relatively narrow compared to the width of the shelf so that it can be easily attached and not be clumsy in appearance.

The design constraints on the part are not too severe since the part is not required to fit in a closely defined space and consequently there is a wide latitude in material and size combinations that can be used to perform the necessary function. In a case like this, economic considerations help in selecting a material and configuration that will perform the support function in the most efficient manner. From a structural standpoint there is only one basic requirement for the part; that it support the shelf and its contents. There are two possible modes of failure of the part. The load can excede the tensile strength of the support and it will break. The alternative possibility is that with the load on the shelf, the support will stretch until the outer edge of the

Fig. 4-1. Typical small thermosetting and thermoplastics molded parts. (*Courtesy Tech Art Plastics Co.*)

shelf tilts down to the point where the contents will slide off. The first is a strength failure and the second is a deformation failure. Both possibilities should be examined, although it is apparent in this case that failure was due to deformation.

The first step in the design process is to tentatively select a material for the part. In this case, we would like a material which is readily

Fig. 4-2. Chairs are matched mold polyester glass products. The walls are high pressure laminated sheets. (*Courtesy Formica Corp.*)

colored for decorative reasons. The shelf may be used in an appliance, in the kitchen, in a chemical laboratory, in a hot place, etc., thus materials should be selected that have the appropriate properties to function in the intended environment. Since the support is intended to maintain a fixed spacing a rigid material with a high elastic modulus is prefered to minimize distortion under load. From the required shape we will assume that the support will be molded, and this should be considered in the selection process. Using the appropriate limitations and a list of plastic materials such as the tables in *Modern Plastics Encyclopedia*, or using a computer selection technique (such as the G.E. service), a group of possibilities is arrived at. The data usually available for such materials are as follows:

1. A tensile strength vs stress curve which is usually done at one of the standard ASTM test rates in the range of 10 inches per minute. From this curve it is possible to determine the yield strength, the ultimate tensile strength, the elastic modulus, and the elongation to failure at the testing rate.

Fig. 4-3. This 420 gal. spray tank was rotationally molded in one piece of polyethylene. (*Courtesy Arnas Molded Plastics*)

2. Creep curves at various stress levels and usually at room temperature (20°) and one elevated temperature, frequently the temperature which the supplier feels is the maximum material use temperature.
3. The glass transition temperature for glassy polymers and the crystalline melt temperature for crystalline polymers.
4. The impact strength both notched and unnotched at room temperature and some lower temperatures such as 0° C or −40° C.
5. Hardness measurements on the Shore or Rockwell Scales.
6. Some derived strength terms such as flexural strength and modulus.

There are various other data which are reported but the above are pertinent to structural design. They will permit us to make some

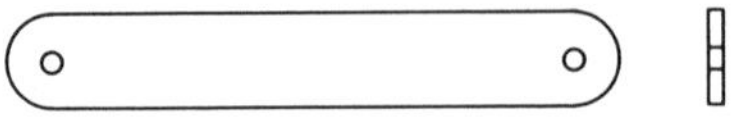

Fig. 4-4. Link for shelf support.

tentative selection of materials and then, after evaluating each, to select a material and configuration that will perform the necessary support function for our part.

Although other materials will obviously be worth consideration, in order to simplify our analysis we will select three materials—polystyrene, polypropylene, and glass-filled polystyrene. The first two are commodity resins which are relatively low cost and the third is a filled material which has much improved stiffness and strength at higher cost. We now have a glassy polymer, a tough crystalline polymer, and a filled glassy polymer. A comparison of these three materials on the basis of strength, creep, and general utility consideration will show the design analysis steps and how to use the data available to make a sound engineering based choice. To completely define the problem it is essential to know the maximum load the shelf must carry (assume 1000 pounds), the time the load is to be carried in long-term static loading (assume 5 years), the maximum deformation that the support can tolerate (assume 5% elongation), the pattern of loading and unloading of the shelf if not continuously used (assume the load-time diagram, Fig. 4-5) and finally some starting dimensions such as the length (10 inches) and the maximum width (1 inch).

Since a shelf is normally used at room temperature we will do the example for room temperature first and then extend the analysis to some elevated temperature.

The first step is to calculate the load on the support. Since the back

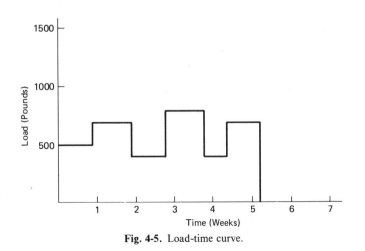

Fig. 4-5. Load-time curve.

of the shelf is supported by the wall hinge, half of the weight of the shelf and its contents will be held by the two supports; each support then carries 250 pounds. The figure of merit used to determine which material should be used will be the cross-sectional area of the support. This figure of merit may be impractically small to mold. It may be less economical for the small section molded from some materials than from another material where the minimum moldable section is adequate to act as a support and the material is cheaper. In this case the structural analysis is useful only in preventing a failure-prone material from being used.

Using the data in Table 4-1 the short-term tensile strength leads to the following calculations of thickness.

$$\text{Load} = \text{tensile strength} \times \text{area}$$
$$\text{Load} = \text{tensile strength} \times \text{width} \times \text{thickness}$$
$$250 = \text{tensile strength} \times 1 \times \text{thickness}$$

for polypropylene the tensile strength is 4500 psi

$$250 = 4500 \times \text{thickness} \qquad \text{thickness} = 0.055 \text{ inches}$$

for polystyrene the tensile strength is 8000 psi

$$250 = 8000 \times \text{thickness} \qquad \text{thickness} = 0.03125 \text{ inches}$$

for glass-filled polystyrene the tensile strength is 12,000 psi

$$250 = 12,000 \times \text{thickness} \qquad \text{thickness equals } 0.0208 \text{ inches}$$

It is apparent that on the basis of the ultimate tensile strength any of the materials would perform at reasonable thicknesses. Applying the criterion of restricting the deformation to 5% we can calculate the apparent modulus at 5000 hours (the initial stress divided into the

Table 4-1. Tensile Strength and Tensile Creep Data.

	Polypropylene	Polystyrene	Glass-filled Polystyrene
Tensile strength	4,500 psi	8,000 psi	12,000 psi
Tensile modulus (short term)	200,000 psi	500,000 psi	1,000,000 psi
Apparent modulus for stress level and 5000 hours at room temperature	37,000 psi 1,550 psi	200,000 psi 200 psi	700,000 psi 10,000 psi
Apparent modulus for stress level and 500 hours at 140°F			250,000 psi 10,000 psi

strain at time) for the three materials. These are tabulated in Table 4-1 along with the initial tensile modulus. The calculations for thickness are as follows:

The stress equals the load divided by the loaded area $= 250/1 \times$ thickness. For polypropylene the initial modulus is 200,000 psi and the apparent modulus at 1550 psi stress and room temperature at time T is 37000 psi.

$$\% \text{ elongation} = \frac{250/t \times 100}{200,000} = 5\%$$

$$\text{stress level} = 250/0.025 = 10,000 \text{ psi}$$

$$t = 0.025$$

$$\% \text{ elongation (5000 hours)} = \frac{250/t \times 100}{37,000} = 5\%$$

$$\text{stress level} = 250/0.135 = 1851 \text{ psi}$$

$$t = 0.135$$

The stress level excedes the assumed stress level for the creep modulus which indicates that the thickness must be increased. Using the stress level for the creep data at 1500 psi the thickness will be $250/1500 = .167$ inch.

For polystyrene the initial modulus is 500,000 psi and the apparent modulus at 200 psi (5000 hours) is 200,000 psi.

$$\% \text{ elongation} = \frac{250/t \times 100}{500,000} = 5\%$$

$$t = 0.01$$

$$\text{stress level} = 250/0.01 = 25,000 \text{ psi}$$

$$\% \text{ elongation (5000 hours)} = \frac{250/t \times 100}{200,000} = 5\%$$

$$t = 0.025$$

$$\text{stress level} = 250/0.025 = 10,000 \text{ psi}$$

This level excedes the design level of 2000 psi based on the data for creep. Using the creep test data as a criterion $t = 250/2000 = 0.125$ inch. For glass-reinforced polystyrene the initial modulus is 1,000,000 psi and the apparent modulus at 10,000 psi (500 hours) is 700,000 psi

$$\% \text{ elongation} = \frac{250/t \times 100}{1,000,000} = 5\%$$

$$t = 0.005$$

$$\text{stress level} = 250/0.005 = 50,000 \text{ psi}$$

$$\% \text{ elongation (500 hours)} = \frac{250/t \times 100}{700,000} = 5\%$$

$$t = 0.007 \text{ inch}$$

$$\text{stress level} = 35,714 \text{ psi}$$

This stress excedes the design stress for the creep data. Using the creep data at a stress level of 10,000 psi the thickness is $250/10,000 = 0.025$ inch.

From this analysis we can see that the determining factor in the design is the maximum stress level that will permit a creep in the range of 5%. Based on this factor, it is apparent that the material to be used, if minimum thickness is the criterion, is the glass-filled polystyrene. From an economic standpoint this is also the best material since the thickness is 20% of the polystyrene and 15% of the polypropylene while the relative cost of the glass-filled material is about 2.5 times that of the polystyrene or polypropylene and the molding costs would be comparable for the three materials.

The shelf is assumed to be loaded and unloaded several times during the design life of 5 years. When the load is removed, the link element recovers partially toward its original length. From our analysis, using the principle of superposition, the recovery would be the elastic strain and a portion of the viscoelastic strain, depending on the length of time allowed for recovery. The creep portion of the strain is permanent.

The curves available for creep are the time extension curves. The complementary relaxation curves are rarely available and would involve a large mass of data since the curves would vary depending on the time that the load was on. Although this is an approximation, the inverse curve is used as a relaxation curve. This assumes that the strain recovery follows the same mechanisms as the straining process which cannot be exactly true as the previous discussion on molecular reaction

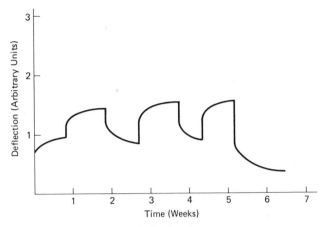

Fig. 4-6. Deflection time curve.

indicated. For engineering purposes, the approximation is sufficiently accurate for most purposes.

Figure 4-6 shows the pattern of long-term strain that corresponds to the stress history of Fig. 4-5 for each of the three materials selected. Each of the curves represents a graphical solution to the strain-stress time for the problem.

To determine the elevated temperature performance of the part, or, alternatively, the dimensions required to operate within the limits imposed, we repeat the calculations. Since the short-term strength is not the limiting factor only the creep analysis is required. The elevated temperature will reduce the strength and increase the creep. Since the glass-reinforced polystyrene is the choice at room temperature, it will be the choice at elevated temperature since it loses strength less with temperature than the unmodified polystyrene and the polypropylene.

At 140° F, 10,000 psi, and 5000 hours the apparent modulus is 250,000 psi.

$$\% \text{ elongation (5000 hours)} = \frac{250/t \times 100}{250,000} = 5$$

$$t = .020$$

$$\text{stress level} = \frac{250}{.020} = 12,500$$

Since this is higher than the assumed level for creep, we use the stress level for creep

$$t = \frac{250}{10,000} = .025 \text{ inch.}$$

To complete the sizing we use the creep curve to calculate the strain and use the stress time profile of Fig. 4-5 to obtain the strain-time curve of Fig. 4-6 for the elevated temperature condition. This represents the graphical solution for this problem.

The last step in the design of the part is to provide a suitable safety factor. In this case twice the section would represent a conservative design allowance so that our glass-reinforced shelf support would have a cross section of 0.50 inch × 1 inch and would safely resist the loading cycle imposed for the 5-year design life of the part.

The simple tension problem includes all of the steps in the structural design problems listed below:

1. Material selection based on functional requirements of the part, plus structural requirements.
2. Evaluation of the relative merit of several potential materials by structural evaluation, by strength requirement, and strain creep at design life criteria.
3. Selection of the material based on the minimum cost, minimum size, temperature limitation by calculation of the shape requirement at the maximum time, and temperature for use.
4. Calculation of the final dimensions based on the stress-time curve for the design life of the part.
5. Adjustment of the dimensions by a suitable safety factor.

A more accurate answer could be obtained directly if the data for a stress-strain time diagram such as Fig. 4-7 were available at a range of temperature. If the curve were available at one temperature with the constants for the WLF equation, it could be generated at the minimum design temperature and the calculation for strain at the design life done directly.

The design can be done in several ways with computer assistance. If the data banks are available on materials with information entered on the apparent modulus at various stress levels and temperatures, the data bank can be interrogated to supply materials with the other required characteristics and then interrogated with a desired

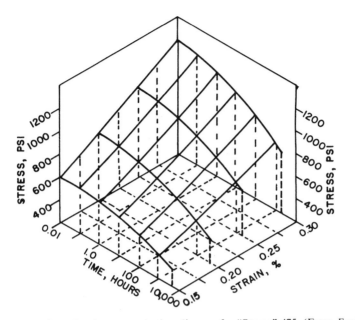

Fig. 4-7. Three-dimensional stress-strain-time diagram for "Styron" 475. (From *Engineering Design for Plastics*, Eric Baer, Reinhold, 1964)

maximum cross section. The materials would be reduced to those that meet the design requirements for the maximum allowed sections. A comparison of the cost effectiveness of these materials would select the optimum material.

The computer can also be used to provide a singular solution to the required cross section. The time-stress-strain data to generate a set of creep curves for a specific material would be provided. The computer is then programmed with the problem of the stress-time profile for the part. Using the inverse curve as the strain relaxation curve, the computer can do an iterative solution to determine the minimum cross section. That would restrict the creep to the set amount in the desired design time. The WLF transformation can be done on the basic stress-strain time data to provide the solution for different operating temperatures.

The problem of compression loading and simple shear loading would be done in the same manner. The only difficulty is that data on compression and shear stress-strain-time relationships are not as readily available as the data on the tensile strain. The Poissons ratio transposition to data from tensile data is comparatively poor in correlation. For an important part analysis it may be necessary to do an accelerated creep test at elevated temperature and use the WLF relationship to estimate the creep values at the necessary operating temperatures.

This chapter gives a problem involving a simple tension member supporting a static load that is analyzed to show the steps involved and how the available data can be used to calculate the size of the part required to take the static load with the expected stress time pattern anticipated.

Chapter 5

Structural Design of Beams, Plates and Other Structural Members

In the previous chapter we dealt with the design of plastics parts to withstand simple structural stresses such as tension, compression, and shear. Most plastics parts are subjected to more complicated load conditions such as bending and buckling stresses. For example, the shelf supported by our hypothetical support member has usually both distributed and concentrated loads on it and the shelf is a beam supported by the shelf hangers as shown in Fig. 5-1.

The bending loads on a beam induce counteracting couples in the material of the beam and these represent the stresses in the beam. The beam stress for elastic materials has been explored extensively in texts such as Timoshenko and the relationship between the bending loads, the stress levels and the deformation of the beams and other members subject to bending stresses are well known. These relationships do not hold for materials that are not strictly elastic in behavior and are quite inaccurate in predicting the behavior of plastics materials which exhibit viscoelasticity and creep.

The best approach to determining the differences in analysis between an elastic beam and one made of a plastics material would be to continue with our example of a shelf; however, this time we will look at the shelf that our previous illustration was supporting. The shelf is an example of a simple beam with several concentrated loads. There are two relationships that apply to define the performance of the part and indicate the probable mode of failure. The first of these is the stress in the extreme outer surface of the beam, or the extreme fiber stress. The second relationship is the deflection of the beam under load. The mathematical expressions for an elastic material are:

$$S = \frac{M : c}{I}$$

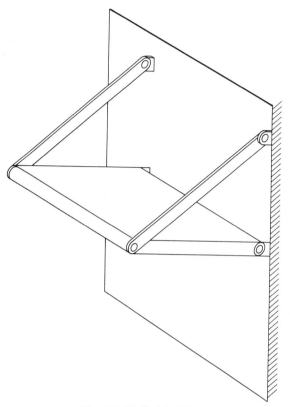

Fig. 5-1. Shelf with link.

where

S = the extreme fiber stress
M = the maximum bending moment in the beam
c = the distance from the neutral plane of the beam to the outer fiber
I = the moment of inertia of the beam cross section

$$d = \frac{Pb(1^2 - b^2)^{3/2}}{9\sqrt{3}\,1\,EI}$$

where

d = maximum deflection due to load
P = the load
I = the length of the beam
b = the distance from the left end of the beam to the load
E = the modulus of elasticity in flexure

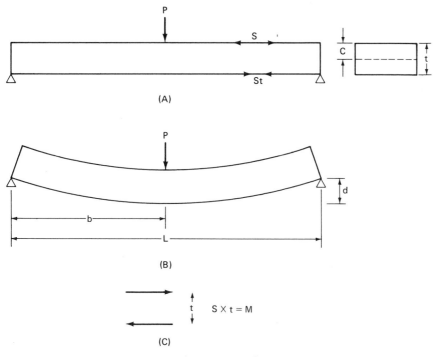

(A)

(B)

$S \times t = M$

(C)

Fig. 5-2. Beam bondings.

Figure 5-2 shows the beam depicting the various quantities as they relate to the mathematical relationship.

The relationship in Eq. 1 is based on the elastic response of the beam material which would produce the stress relationship shown in Fig. 5-3A. The couple resisting the bending moment is made up of the elastic response of elements of the material alternately in tension and compression above and below the neutral plane. The material is assumed to have the same modulus in tension and in compression so that the neutral plane is in the center of the beam. The stress level is proportional to the distance from the neutral plane and is a maximum at the surface of the beam.

It is important to see how this relationship is altered by introducing some material property changes. For example, if the moduli of elasticity in tension and compression are different, the neutral plane is displaced from the center of the beam and the stress distribution is altered as shown in Fig. 5-3B. The relationship for the maximum fiber stress still applies if we take into account the new location for the neutral plane and the fact that the modulus is different in tension and

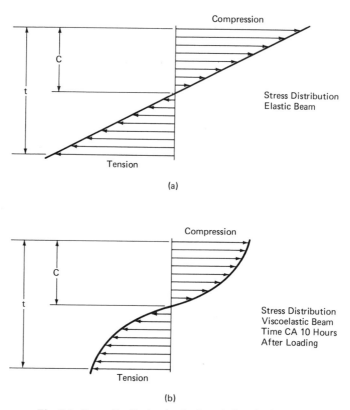

Fig. 5-3. Stress distribution in elastic and viscoelastic beams.

compression. It is apparent that the moment of inertia is now the sum of the moments of the rectangles below and above the neutral axis and this is different from the moment of inertia of the complete rectangle about its central axis. In addition, the effect of the different moduli is to increase the deflection of the beam in the direction of the lower modulus direction. Fortunately for the mathematical treatment, the difference in the tensile and compressive moduli is small compared with the absolute value of the modulus so that the mid-plane treatment can be used with relatively little error, especially since the outer layers of the beam take the majority of the stress.

The value of the elastic modulus in bending, which is taken as the effective average of the tensile and compressive moduli (usually measured directly), is a function of the loading rate and time as are the compressive and tensile moduli which are its components. As a result, the deflection of a beam is a time-dependent function. The initial deflection of the beam will be that related to the initial modulus of the material or the slope of the rapid load stress-strain curve. As the material undergoes creep and stress relaxation, the stress distribution through the section of the beam undergoes a rearrangement. Since the apparent modulus of elasticity is a function of time and stress level, the value of the modulus decreases more in the highly stressed skin portions of the beam and less in the inner layers of the beam. As a result the triangular stress distribution which exists for an elastic material beam decays into the trapezoidal distribution shown in Fig. 5-4. This is an idealized distribution, assuming that the change of modulus is proportional to the stress level. The relationship is usually more complex and a curved stress distribution is more likely such as shown in Fig. 5-3B. Calculation of the effects of time on bending is approximated by a trapezoidal stress distribution.

From a functional standpoint what happens is that the beam will continue to deflect with time. If the extreme fiber stress does not exceed the tensile yield strength of the material this deflection will proceed at a decreasing rate but never become stationary in value. The design problem is usually one of distortion. When the beam finally bends to the point where it is no longer functional, this is the design of the plastic beam. In the case of the shelf that we are considering as our example, the failure point will be reached when the shelf has sagged to the point where the object placed on it will slide to the low point. Other plastics beams would have other failure points as a result of creep and stress relaxation. A beam structure (Fig. 5-5) in a machine housing would be considered to have failed by excessive deflection if it moves to the point where it interferes with the mechanism it enclosed or if it sagged to the point where it no longer acts as an effective enclosure. A plastics roof section (Fig. 5-6) may be considered to have failed if it sags to the point where it looks unsafe or ugly, or it may sag to the point where it interferes with the normal use of the space it covers.

The decision as to what is considered excessive deflection is made based on product or structure requirements. Once it has been found how much alteration of the initial geometry of the structure can be permitted and still have it useful, we can translate this into the allowed deflection of the beam at the design life under the imposed loads. The solution to this problem is not easy since it is necessary to estimate the

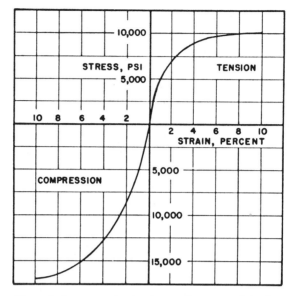

Stress-strain curve in tension and compression for polyoxymethylene.

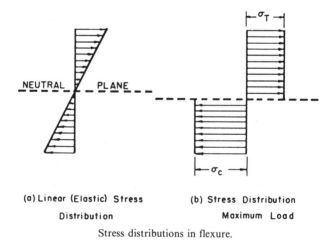

(a) Linear (Elastic) Stress Distribution

(b) Stress Distribution Maximum Load

Stress distributions in flexure.

Fig. 5-4. (From *Engineering Design for Plastics*, Eric Baer, Reinhold Publishing, New York, 1964.)

necessary beam cross section and compute the initial stress levels caused by the loads. From these stress levels and the design life of the beam it is possible to find the apparent modulus of elasticity at T_d the design life. This value can be put into the elastic deflection equation

Fig. 5-5. Chlorine pulp washer hoods spanning more than 20 feet fabricated of polyester resin for maximum corrosion resistance. All structural members, as well as nuts and bolts, are of FRP construction. (*Courtesy Durez Div., Hooker Chemical Co.*)

to give the deflection at time T_d. If the deflection is excessive the beam is increased in section, the new stress levels are calculated, and a new apparent modulus determined. By this process it is possible to arrive at the proper beam dimension to deflect the design amount at the design life of the beam. This type of calculation can be programmed for computer solution and such solutions have been done.

One of the requirements for solution of this problem is the availability of data on the apparent modulus of elasticity as a function of time stress level and temperature. Since this is a derived quantity based on the creep curves good creep data at several temperatures and stress levels are required. This can be used in conjunction with the WLF relation to generate a complete set of creep curves that can then be converted to apparent modulus curves, or tables, which can be used

Fig. 5-6. Fiber glass panels of reinforced plastics material are used for the roof of this large warehouse. (*Courtesy International Filon Producers Assn.*)

in the computer solution to the bending equations. An equation of state for a material would of course be much better in that it would give an exact relationship between stress, strain, time, and temperature but these have not been readily formulated for plastics materials, as has been previously discussed. If such an equation of state is available either in explicit form or as a set of data points that can have a curve fitted, it is used by setting the limiting temperature for use and solving for the stress-strain-time relationship. The derivative of the equation d stress$/d$ strain as a function of time and stress level is the apparent modulus curve except that this would give an exact instead of an approximate value for the modulus. If such an equation is available, it would be possible to sum the deflections with respect to time and, given a specific initial geometry and load, determine the design life. The solution would then involve a series of calculations to bring the design life up to the desired value with the allowed deflection as a constant. This calculation could also be programmed as a computer operation.

A sample calculation is given below:

SHELF PROBLEM CALCULATIONS*

The shelf problem assumes a bookshelf with a uniformly distributed load of 250 pounds on a shelf 48 inches long by 12 inches wide. The problem is to find the thickness of the shelf in an appropriate material which will limit the sag of the shelf to 2 inches maximum for a useful life of 4 years. The first material examined is polypropylene homopolymer with an initial elastic modulus of 170,000 psi and an apparent modulus vs time as shown in Fig. 5.9. The shelf problem is shown in Fig. 5-7.

From Roark formulas for stress and strain P 106 we have:

$$y_{max} = sag = \delta = \frac{5\,wl^2}{384\,EI} \tag{1}$$

$$I = \frac{1}{12}\,ah^3 \tag{2}$$

where

I = moment of inertia
a = width of shelf
l = length of shelf
E = modulus of elasticity
h = thickness of beam (shelf)
w = total load

$$I = \frac{1}{12} \cdot 12 \cdot h^3 = h^3$$

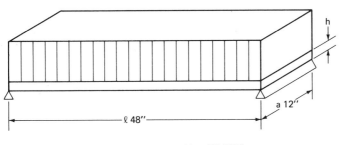

Distributed Load W 250#

Fig. 5-7. Illustration for shelf problem.

*A basic understanding of computer technology is essential to a complete understanding of these calculations.

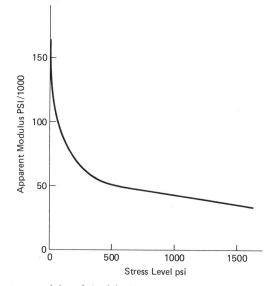

Fig. 5-8. Apparent modulus of elasticity level at 4 years polypropylene homo polymer.

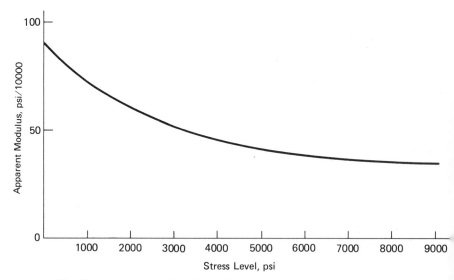

Fig. 5-9. Apparent modulus of elasticity polypropylene 30% glass at 4 years.

substituting into Eq. (1)

$$\delta = \frac{5 \cdot 250 \cdot \overline{48}^2}{384 \cdot 170000 \, h^3} = 2 \text{ inch}$$

$$h^3 = \frac{5 \cdot 250 \cdot \overline{48}^2}{384 \cdot 170000 \cdot 2}$$

$$h = \sqrt[3]{\frac{5 \cdot 250 \cdot \overline{48}^2}{384 \cdot 170000 \cdot 2}} = 1.019 \text{ inch}$$

This is the solution. Assuming no creep

$$S_m = \text{maximum stress} = \frac{Mc}{I}$$

where

$M =$ maximum bending moment
$c =$ half the thickness $= n/2$

From Roark

$$M \max = \frac{wl}{8}$$

$$Sm = \frac{\dfrac{250 \cdot 48}{6} \cdot \dfrac{1.019}{2}}{1.019^3} = 714 \text{ psi}$$

From the apparent modulus curve Fig. 5-8 the apparent modulus for 4 years and a stress level of 714 psi is 48,000 psi. The first creep approximate solution is therefore:

$$h = \sqrt[3]{\frac{5 \cdot 250 \cdot \overline{48}^2}{384 \cdot 2 \cdot 48000}} = 1.553 \text{ inches}$$

The corresponding maximum stress is:

$$Sm = \frac{\dfrac{250 \cdot 48}{6} \cdot \dfrac{1.553}{2}}{1.553^3} = 310.18$$

For a stress level of 310.8 psi and 4 years the apparent modulus is 96,000 psi. Using this we recalculate the thickness h

$$h = \sqrt[3]{\frac{5 \cdot 250 \cdot \overline{48}^2}{384 \cdot 2 \cdot 96000}} = 1.233 \text{ inches}$$

$$Sm = \frac{\dfrac{250 \cdot 48}{6} \cdot \dfrac{1.233}{2}}{1.233^3} = 493.6 \text{ psi}$$

The apparent modulus for 4 years and 493.6 psi is 51,000 psi. From this information it is evident that the solution lies between 1.553 and 1.233 inches. By trial and error it can be determined that the actual thickness is 1.38 inches, for which the corresponding actual sag at 4 years is 2.105 inches. The various values of h are assumed between 1.5 and 1.25 inches and the sag calculated for each value by determining the stress level and using the corresponding apparent modulus from Fig. 5-9.

It is apparent that this type of computation is a good computer operation. A flow chart and computer program to do this follows in Figs. 5-10 and 5-11.

It is instructive to solve the problem for another material with a much higher modulus. The problem is redone using the values for glass-filled polypropylene. The initial modulus of elasticity is 900,000 psi.

$$h = \sqrt[3]{\frac{5 \cdot 250 \cdot 48^2}{384 \cdot 2 \cdot 900000}} = 737$$

$$S_{max} = \frac{\dfrac{250 \cdot 48}{6} \cdot \dfrac{.737}{2}}{.737^3} = 1381.5 \text{ psi}$$

At 4 years and a stress level of 1381 psi the apparent modulus is 670,000 psi for glass-filled polypropylene according to the curve in Fig. 5-9. The new solution for the creep case first approximation is:

$$h = \sqrt[3]{\frac{5 \cdot 250 \cdot 48^2}{384 \cdot 2 \cdot 670000}} = .64$$

$$S_{max} = \frac{\dfrac{250 \cdot 48}{6} \cdot \dfrac{.64}{2}}{.64^3} = 1786 \text{ psi}$$

The modulus for a stress level of 1786 psi at 4 years is 620,000 psi and the corresponding h is .659 inch. It is apparent that by recycling or by the use of the computer program and a computer the answer can be accurately calculated. The main point of present interest is that the value for the glass-filled material is approximately one half of the value for the unfilled material.

It is interesting to note some of the results of our calculations on the shelf. The value (assuming an elastic response in several materials) is tabulated, as is the value for the design life and deflection, and a

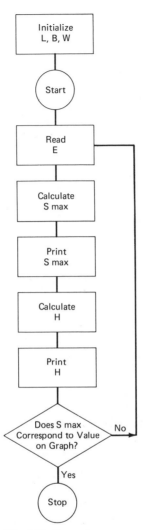

Fig. 5-10. Computer sequence.

value which is frequently used, of the elastic value with a safety factor of three. It is apparent that depending on what the plastics is, the three values diverge drastically from each other to an extent that involves either an extreme penalty in cost, a likely failure or—for the carefully designed—a bonus of low cost and good life of the part.

Probably the greatest deterrent to the implementation of this design approach—which is admittedly not exact—is the lack of data on which

```
        ∇SHELF[□]∇
     ∇ SHELF
[1]  LOOP:'ENTER APPARENT MODULUS'
[2]    E←□
[3]    'MAXIMUM STRESS IS ';(4.1039÷L)×((W×E*2)÷B)*0.3333;' PSI'
[4]    'SHELF THICKNESS IS ';0.4275×L×(W÷(E×B))*0.3333
[5]    →LOOP
     ∇


     L←48
     B←12
     W←250


     SHELF
ENTER APPARENT MODULUS
□:
     170000
MAXIMUM STRESS IS 721.3 PSI
SHELF THICKNESS IS 1.0196
ENTER APPARENT MODULUS
□:
     48000
MAXIMUM STRESS IS 310.47 PSI
SHELF THICKNESS IS 1.554
ENTER APPARENT MODULUS
□:
     96000
MAXIMUM STRESS IS 492.81 PSI
SHELF THICKNESS IS 1.2335
ENTER APPARENT MODULUS
□:
     →
```

Fig. 5-11. Computer program.

to base the calculations. The data from which the information was taken for the example given was extrapolated from what was available on the three materials examined. One of the important tasks for plastics engineers is developing efficient standardized tests for creep and stress relaxation which can be extrapolated by means of the WLF relationship and similar equation of state correlations to provide the basis for accurate calculations. With the increasing use of polymer-based materials (Figs. 5-12 and 5-13) this is becoming a critical problem. Just as part of the cost of steel, concrete, and other materials is in the characterization of the materials so that they can be intelligently used, it is imperative that this be done for plastics. It is costly data to acquire compared to simple tensile tests but from our example it is obvious that the costs of not having the data far outweigh the cost of obtaining it.

Fig. 5-12. Plastics boats are subject to large and variable stresses.

The analysis of structural members such as struts and columns is more complicated than the simple bending example. If the column is large in area so that there are no buckling stresses, the analysis can procede assuming simple compression loading as was done in the previous chapter. When there are buckling stresses the problem must

Fig. 5-13. Photograph shows excellent condition of first Corvette fiber glass body. Production body.

be analyzed from several standpoints to determine if there will be a buckling failure. First, the analysis must be done, assuming an initial modulus of the material of the strut, to see if the instability criteria for the buckling is exceeded. If the strut is initially stable the next problem is to determine what effect the creep and stress relaxation will have on the strut. As in the case of the beam, there will be a redistribution of the stresses in the part with time depending on stress level and temperature. This will usually lead to more bending of the strut or column and, if the part is near the instability point, lead to buckling instability and failure. There will also be compressive creep in the part which may cancel some or all of the increased bending tendency so that for a time the changes will be such that the strut or column will remain stable.

The design approach is as follows: design the part to have sufficient strength and stiffness using initial modulus values and the standard formulas for buckling, determine the change in modulus at the stress level and temperature for the application at a series of different increasing times, and determine the time at which the instability limit has been exceded. The beam or strut dimensions are then changed to reflect greater initial stability and the process is repeated to find the time at which instability occurs. By an iterative process the strut dimensions are changed and failure times determined until the time equals or somewhat exceeds the design life requirements for the part. As in the case of the beam in bending, this can be programmed for computer calculation.

Buckling instability occurs with edgewise loading of a sandwich panel. Since the skins take most of the load, it is essential that they are securely adhered to the core material which has virtually no compressive resistance. Failure by buckling can be one of two. The skins can accordion-fold due to the compressive loading without the bond to the core failing. This type of failure reflects the highest strength of the sandwich panel in edge compression loading. The core is crushed and the skins adhere to it to give maximum strength until the supported skin fails. If the bond to the core material fails first, the skins buckle away at very low edge compression loads to produce weak products. The usual age problems are involved with creep in the adhesive joint between the skins and the core. In the event that this bond is made directly by the core material (as is the case in urethane foam core panels) the creep characteristics of urethane resin material requires evaluation. In the event that the bond is established using a separate adhesive (as might be used in a styrene foam core panel) it is necessary that an evaluation of the bond to the core material be made to determine the bond between the skin material and the styrene foam. A failure of the bond at any of the interfaces will result in the premature buckling failure of the panel in edge compression.

The design life consideration is then an evaluation of the change of bond strength with time as a result of creep and stress relaxation. This is one area where it is difficult to obtain good analytical solutions because of the complex stress patterns in the adhesive joint. The process of joining produces a substantial amount of residual stress in the joint, the directions of which are generally unknown. Thermal stresses caused by the joining process add further to the problem. In the case of practical engineering, it is advisable to test panel combinations to insure that the surface bond is more than adequate to cause accordion buckling and then limit the loads to about 50% of the load to cause accordion failure.

There are two other types of bending problems that are common. One is bending in a plate where the plate is supported around the edges and the bending takes place in more than one direction. An example of this would be the bending that occurs in the header at the end of a pressure tank or the bending in a continuous plate such as the floor of an enclosure. In each case there is a formula for the solution of the bending deflection in terms of the modulus of elasticity and the moment of inertia of the cross section. Roark's book on *Formulas For Stress and Strain* contains a complilation of formulas on different types of structural elements under bending conditions. It uses the same approach as used in the simple beam case; namely, first doing the elastic solution for the beam, then determining the time-dependent bending effect and redesigning the beam geometry based on this, until the deflection is within design limits at the design life of the structure. An iterative calculation such as this gives the best available engineering solution for most materials. Another approach is given in Chapter 4 of *Engineering Design for Plastics* but the required input data are even less available for this type of solution than for the simpler one using the apparent modulus. The accuracy of the results is not enough better in most instances to justify the detailed calculations. The major improvement in design comes from the basic step of including the effect of the time-dependent strain increase.

The other type of bending problem is one related to cyclical loading of the bending element. This would be a long time load which is removed for a length of time and then replaced as, for example, if our shelf had objects added and then removed from time to time. In some cases taking this condition into account will result in a much lighter structure than would result from assuming that the maximum possible load is applied for the design life of the part. An example that may be given is the use of a beam to support traffic—a floor for a dwelling or a vehicle. Loads may be applied for periods ranging from hours to days, to months, to indefinitely. Assuming a normal pattern of use, it is possible to estimate the average load time burden of the structure and to calculate the recovery that takes place between loading times. While it is not strictly accurate to assume that the recovery curve follows the same curve as the initial viscoelastic deformation, for practical purposes this is the only data that are available and will be a good estimate. The creep curve is inverted to form a strain recovery curve and based on this curve, the recovery deflection can be estimated. The calculation involves estimation of the initial recovery modulus (elastic behavior) and the viscoelastic recovery modulus which would be related to the strain recovery divided by the strain recovery time. The redistribution of the stress in the beam section would have to be estimated in order to determine the proper EI product to substitute in the bending deflection formula to calculate the recovery. This approach is obviously much more useful for loads which are applied and removed in times of hours to days since the creep curves can be more easily worked within these regions by a curve fitting procedure. When it is apparent that large savings in material and structure weight can be effected the calculations would be justified.

One general point should be made. Most of the formulas developed for bending deflection are based on relatively small strain levels compared with the geometry of the structures. The mathematics do not apply to large strains and the assumptions of linearity of material behavior do not apply at large strains. (Fig. 5-4) When the solutions of bending problems show large strains, they should be checked by experiment or much larger safety factors invoked. An alternative approach is to substitute materials which have higher moduli and lower ranges of creep for materials that show large strains at the design life required. Most plastics structural applications require materials that have glass and other high modulus reinforcements. The degree of confidence in the result of the analysis is much higher and the probability of failure is lower.

This chapter has reviewed the methods of designing parts to take static bending loads. The approach covered involves the use of the elastic bending relationships adjusted to take into account the effects of creep by the use of an apparent modulus of elasticity derived from creep data. The method gives a good approximation to the behavior of plastics members under sustained loading. The accuracy is as good as the available data. More rigorous solutions are available that require relationships for the time dependent modulus of elasticity of the materials which are usually not available nor derivable from the type of creep data available on commercial plastic materials. The approximate solutions are a substantial improvement over those which neglect the creep of plastics materials and the probability of failure by excessive deflection as a result of creep. The initiation of catastrophic failure by creep was discussed for end loaded columns, struts, and structural sandwich panels and a design approach suggested for this condition.

Chapter 6

Dynamic Load Response of Plastics Members and Effects of Cyclical Loading

In the previous discussions on loading of plastics members, the concern was with loads applied at a constant level for relatively long times, in the order of hours or longer. The present discussion relates to loads applied to plastics for very short periods of time in terms of seconds or less. Since plastics materials exhibit a complicated stress response we would expect that the behavior under this type of load would differ substantially from the behavior under static loading, and it does.

Under short time or cyclical loading, most plastics materials do not exhibit significant creep unless they are stressed near their breaking strength. In general, the response is viscoelastic and the modulus of elasticity tends to be close to the values obtained in conventional testing at high speeds. One important fact to note, though, it that typically a material exhibiting viscoelastic behavior does not follow the same stress–strain curve when the load is removed as it does when the load is applied. Figure 6-1 shows a curve of loading and unloading for a typical viscoelastic material with the curves for several peak stress levels. There are two important things to read from these curves. First the area between the two curves, increasing load and decreasing load, is a product of force and distance and it represents work energy expended on the material that is not recovered. Consequently, it must appear in the material as frictional heat caused by viscous drag between the molecules. The second factor to note is that higher stress levels lead to significantly larger areas contained between the loading and unloading curves more than proportional to the increased load. The area between the curves is referred to as a hysteresis loop to represent the energy loss by viscous drag in the structure.

Figure 6-2 shows the relationship between the value of the short time modulus of elasticity and temperature for a typical viscoelastic

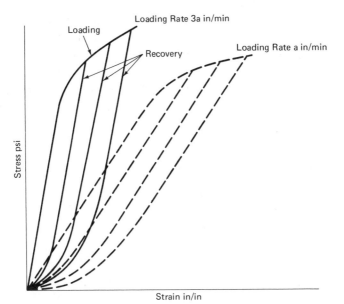

Fig. 6-1. Stress strain curves with recovery curves for typical viscoelastic material.

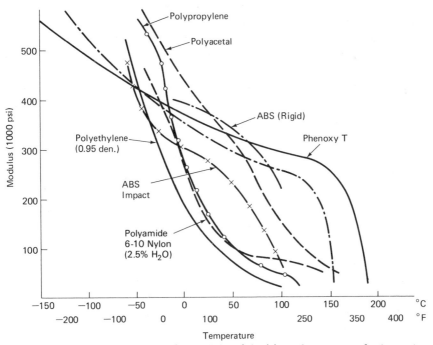

Fig. 6-2. Relationship between short time modulus of elasticity and temperature for 6 - 10 nylon.

plastics material 6-10 nylon. This relationship shows that the modulus drops rapidly with increasing temperature and undergoes some rather severe drops at the glass transition temperature and the crystalline melting temperature. In view of this temperature dependency of the modulus and the fact that cyclical loading generates heat that will raise the temperature of a plastics material, it is apparent that this situation requires attention in designing parts to withstand cyclical loads.

Repeated application of stress acts in another way on materials. The effects of stress in different regions of the material cause the generation of local loads sufficient to cause bond breakage. In addition, the energy inputs created by repeated loading cause phase changes in the material by recrystallization of crystalline polymers and the breakage of bonds at the fold points in crystallites of material. The accumulation of these defects acts in a manner similar to work hardening effects in a crystalline metallic structure to generate structural weakness and eventual failure by fatigue. There are two general modes of potential failure in dynamically loaded plastics parts, one of which can be designed around rather easily and the other of which is a statistical failure condition that can only be dealt with by knowing the cycles to failure of the material at specific stress levels and selecting a level that will have a low probability of failure for the design life of the part.

The hysteresis heating failure occurs more commonly in plastics members subject to dynamic loading. It would be well to point out what comprises dynamic loading and what types of stresses are encountered. One example commonly encountered is a plastics gear. In the course of operation the gear teeth are periodically, once per revolution, subjected to a bending load that transmits the power from one gear to another. Another example is a link that is used to move a paper sheet in a copier or in an accounting machine from one operation to the next. The load may be simple tensile or compressive stresses, but more commonly it is a bending load. There are some less obvious but quite important dynamic stress situations which illustrate the importance of dynamic loading. A belt which is used to drive a pulley is subject to repeated bending and tensile stresses during operation. A tire on a vehicle is subjected to a complicated combination of bending, bearing, and compressive stresses during the movement of the vehicle that it supports. The keys on a typewriter or printer

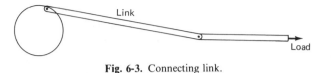

Fig. 6-3. Connecting link.

are subjected to repeated impulse loading during use even though the action is not strictly cyclical in nature. Add to this the effects of vibration induced by vehicle motion or machine action which is an induced cyclical stress in parts which are attached to the vibration-inducing object and it becomes apparent that cyclical loading is a widely encountered type of loading. In many instances the dynamic stress exists in conjunction with static stresses and with other longer term periodic loads.

An example will be given to show how dynamic loading can lead to part failure by hysteresis heating of the part. When this condition exists the failure will be catastrophic rather than gradual. This is not generally true of creep failure or of normal fatigue failure. Our example is a link which is loaded in tension and is the connecting link that drives a flywheel type of unit by means of a linear actuating force (Fig. 6-3). The primary load on the part is alternate tension and compression loading of the part. The compression loading is low since it represents the flywheel driving the link back against friction forces. The tensile force is the driving force and it varies from zero to a maximum which is determined by the torque load on the flywheel member. The relationship of the force to time is shown in the formula in Fig. 6-4 which gives the force function per revolution and this divided by the time of a revolution will give the force as a function

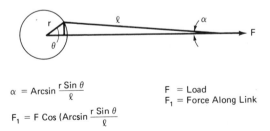

$$\alpha = \text{Arcsin } \frac{r \, \text{Sin } \theta}{\ell}$$

$$F_1 = F \, \text{Cos } (\text{Arcsin } \frac{r \, \text{Sin } \theta}{\ell}$$

F = Load
F_1 = Force Along Link

Fig. 6-4. Force/time relationship for the link shown in Fig. 6-3.

of time. Increasing temperature increases heat transfer from the part to the surroundings and if the rate of heat transfer equals the rate of heat generation at a temperature below the softening temperature of the material, the process will stabilize and the part will not fail. It is apparent that the major factor in hysteresis failure is the ability of the part to dissipate the heat generated to its surroundings. The designer can use several approaches to prevent hysteresis failure. The first is material selection. The stiffer the material is, the smaller the strain is for a given stress level and the lower the hysteresis loss per cycle. Some materials are additionally fairly linear in stress-strain characteristics and have smaller hysteresis loops. These would be preferred in dynamic loading applications. Table 6-1 gives some examples of materials that are used in dynamic load applications because of low hysteresis effects. Figure 6-5 illustrates a cam that is similarly loaded.

The second approach the designer can use is to improve the heat transfer conditions from the part. This can be accomplished in several ways. One way is to operate the part in coolant medium which would also act as the lubricant for the system. The heat transfer to a liquid is usually much better than to air, and the liquid can be cooled by passing it through a heat exchanger device. A second approach is to improve the heat transfer to air. This can be done by increasing the surface area of the part by means of fins or other surface projections. The larger area will increase the heat flow out of the part substantially. The third approach is the use of air circulating techniques which can be areas added to the stressed unit such as air deflector sections or the use of fan cooling as part of the system. Another approach that may fit some applications is the use of metal heat-sink elements buried in the plastic part which conduct the heat into other parts of the complete machine to dissipate it to the surroundings.

**Table 6-1. Materials Having Low
Hysteresis Loss in Cyclical Loading.**

Glass filled polystyrene
Glass filled polyester
Graphite fiber filled nylon
Glass filled ABS materials
Glass filled polycarbonate
Silica filled epoxy

Fig. 6-5. Plastics cam in which cam track has been machined in molded blank with central insert. This cam is made from a graphite–linen phenolic laminate.

Basically, anything that can be done to reduce the temperature of the part by removal of heat generated by the cyclical stress will improve the possibilities of a part surviving the cyclical stress. If the heat transfer capability is limited, then the only alternative is to use stiff materials and low stress levels on the part compared with the strength capability of the material. The heavier parts that result will be relatively inefficient in the use of material. In some cases when the load applied is an inertial load (such as an impeller on a pump) it may be that only a trade-off of weight for low stress level with weight which increases the inertial stress must be made to get an operating unit. Figures 6-6 a, b, c, and d show some of the heat transfer expedients used to improve the performance of parts under cyclical loadings.

To review, the design of a part to withstand cyclical loading is first based on the ability to minimize heat buildup caused by hysteresis. This is done by minimizing the size of the hysteresis energy loss per load cycle using stiff materials and relatively low stress levels. The part is then designed into the operating environment to maximize the heat transfer to the surroundings. The part reaches an equilibrium tem-

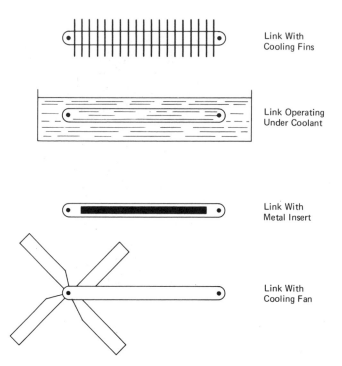

Link With
Cooling Fins

Link Operating
Under Coolant

Link With
Metal Insert

Link With
Cooling Fan

Fig. 6-6. Designs for improved heat transfer from cyclically loaded link.

perature when the internally generated heat is equal to the heat transferred. If this occurs substantially below the softening point of the
material the part will function indefinitely. At this point the damage
done by the cyclical loading is fatigue damage.

Fatigue is the gradual loss of integrity of a material because of the
generation of micro-defects in the structure. In the case of polymer-
based materials these are usually bond fractures and changes in the
structure of crystallites of the material. The effect is generally a loss
of toughness, lowered impact strength, and lowered tensile elongation. The effect is produced because of the microvoids formed when
the structure of the material is altered by the repeated stressing. The
standard way to deal with fatigue effects is by a statistical approach.
A curve of stress to failure versus the number of cycles to this stress
level to cause failure is made by testing a large number of representa-
tive samples of the material under cyclical stress, each one at a progressively lowered stress level. This curve, called an S-N curve, is used

in designing for fatigue failure by determining the allowed stress level for the number of stress cycles anticipated for the part. In the case of materials such as metals, this approach is relatively uncomplicated. Unfortunately, in the case of plastics the loading rate, the repetition rate, and the temperature all have a substantial effect on the S-N curve, and it is important that the appropriate data be used. There are some theoretical relationships that can be used to relate the stressing rate to the endurance of the material given in *Engineering Design for Plastics* (pp. 388–392). In general it is desirable to have specific data generated by accelerated testing to predict the performance of a material in fatigue. Short-term physical properties have little relationship to the fatigue strength.

While the hysteresis effect is a limiting situation in the design of dynamically loaded plastics parts, it is useful in other types of designs. As indicated, one of the serious dynamic loading problems frequently encountered in machines and vehicles is vibration-induced deflection. Such effects can be highly destructive, particularly if a part resonates at one of the driving vibration frequencies. One of the best ways to reduce and, in many cases, eliminate vibration problems is by the use of viscoelastic materials. Some materials such as silicone elastomers, flexible vinyl compounds of specific formulations, urethane plastics, and a number of others have very large hysteresis effects. By designing them into the structure it is possible to have the viscoelastic material absorb enough of the vibration inducing energy and convert it to heat so that the structure is highly damped and will not vibrate. Figure 6-7 shows several examples of damping type combinations. Figure 6-8 shows a viscoelastic liner for a missile launch tube. In each case the viscoelastic material is arranged in such a way that movement or flexing of the part results in large defelections of the viscoelastic materials so that a large hysteresis curve is generated with a large amount of energy dissipated per cycle. By calculating the energy to heat it is possible to determine the vibration levels to which the structure can be exposed and still exhibit critical damping. There is one area that must be evaluated. Polymer materials exhibit a spectrum of response to stress and there are certain straining rates that the material will react to almost elastically. If this characteristic response corresponds to a frequency to which the structure is exposed the damping effect is minimal and the structure may be destroyed. In order to avoid the possibility of this occurring, it is desirable to

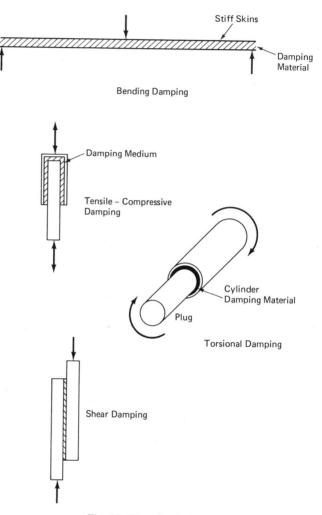

Fig. 6-7. Damping designs.

have a curve of energy absorption vs frequency for the material that will be used. If a low absorption anomoly occurs in the range of application a different material formulation should be used. Figure 6-9 depicts the testing of a railroad tie plate using polyethylene as a damping medium.

The same approach can be used in designing power transmitting units such as belts. In most applications it is desirable that the belts be elastic and stiff enough to minimize heat buildup and to minimize

Fig. 6-8. A section of the viscoelastic compensating liner designed for the Poseidon Missile Launch System. This liner is adhesive bonded to the inner surface of the launcher tube providing an interface to the missile. Polyurethane is the material. (*Courtesy Westinghouse Research and Development Center*)

power loss in the belts. In the case of a driver which might be called "noisy" in that there are a lot of erratic pulse drive forces present, such as an impulse operated drive, it is desirable to remove this noise by damping out the impulse and get a smooth power curve. This is easily done using a viscoelastic belt that will absorb the high rate load pulses. The same approach can be used by making one gear in a gear train or one link in a linear drive mechanism an energy absorber. The viscoelastic damping can be a valuable tool for the designer to cope with impulse loading that is undesirable and potentially destructive to the product.

There is one other type of application where the damping effect of plastic structures can be used to advantage. It has a long although not obvious history. The early airplanes used doped fabric as the covering for wings and other aerodynamic surfaces. The dope was cellulose nitrate and later cellulose acetate which is a damping type of resin material. Consequently, surface flutter was a rare occurrence.

Fig. 6-9. Laboratory testing of tieplates for railroad applications. This project failed in the final field tests.

It became a serious problem when aluminum replaced the fabric because of the high elasticity of the metal surfaces. The aerodynamic forces acting on the thin metal coverings can easily induce flutter and this results in difficult design problems for minimizing the effect. The recent introduction of an all-plastics airplane indicates that this plastics property is again in use. Figure 6-10 illustrates a critical use of boron-epoxy plastics in a helicopter rotor drive shaft. More important is the use of plastics for the outer coverings of structures exposed to severe wind loads and to high speed ground vehicles. The antiflutter potential of plastics skins should be extremely helpful in designing and making elements for these applications.

Fig. 6-10. Bell Helicopter Company engineers hold an 8-foot length of boron/epoxy shafting which will comprise one of three sections that make up the final helicopter tail rotor driveshaft system. An aluminum shaft system would require five shorter sections, with additional end fittings and bearing supports, to transmit an equal amount of power over an equal distance. The lighter weight and superior stiffness of the composite material thus permits an overall system weight reduction of 30%. (*Courtesy of Textron's Bell Helicopter Co.*)

This chapter has covered the reaction of plastics to varying loads. The heating of plastics parts by hysteresis and the methods of designing to avoid frictional heat failures of such parts has been discussed. Fatigue effects on plastics subject to periodic loading as a failure mechanism was reviewed with emphasis on the design engineering approach to avoid fatigue failure. The treatment is concluded with a discussion of the use of viscoelastic effects to damp vibration and transient loads to reduce flutter and, in general, to absorb mechanical energy and prevent damage from induced mechanical vibration. By using the energy absorbing properties of plastics the range of useful components is increased.

Chapter 7

Other Forms of Stress Applied to Plastics Parts

In addition to the static and dynamic loads applied to plastics parts described in the preceding chapters, there are a number of other forms of energy impingement on machine and structural components to which plastics react in a manner strikingly different from other materials. In order to design plastic parts intelligently it is necessary to know the difference in response and degree of response to these other forms of stress.

The significant stresses that will be discussed are as follows:

1. Impact loading
2. Impulse loading
3. Puncture stresses
4. Frictional loads
5. Erosion stress (wind and sand or water)
6. Hydrostatic loads

Impact loads are a particularly important kind of load for plastics. While many materials such as polyethylene and nylon have good impact strength, other plastics such as crystal polystyrene and some grades of polyvinyl chloride have low impact strength. Many of the tests for impact strength have been based on tests for steel and other metals and the applicability of such tests to plastics has always been questionable (Figure 7-1). For example, polyvinyl chloride polymer which rates low in notched Izod impact tests performs comparatively well in normal applications that involve impact loading. On the other hand, some grades of rubber-modified high impact styrenes that show up well in the Izod test break on impact under field test conditions. These results have led to reexamination of the tests used to determine the toughness of plastics. Figures 7-1 F and G show the set up for the test which is generally considered to be much

(A) Tensile test apparatus

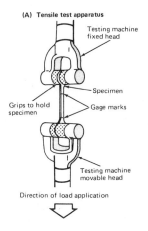

Testing machine fixed head

Specimen

Grips to hold specimen

Gage marks

Testing machine movable head

Direction of load application

(B) Flexural properties test setup

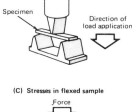

Specimen

Direction of load application

(C) Stresses in flexed sample

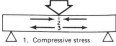

Force

1. Compressive stress
2. Zero stress
3. Tensile stress

(D) Mechanical system of stiffness tester

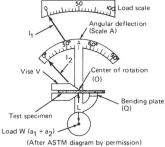

Load scale

Angular deflection (Scale A)

I_1

I_2

Vise V

Center of rotation (O)

Bending plate (Q)

Test specimen

L

Load W ($a_1 + a_2$)

(After ASTM diagram by permission)

(E) Stiffness-in-flexure apparatus

Load indicator

Angle indicator

Specimen clamp

Specimen

Crank for initial loading (manual)

Rod to add weights (needed to test stiffer specimens)

(Free end of specimen impinges on this projection, causing disk carrying load indicator to move.)

(F) Izod impact specimen

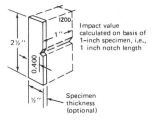

IZOD

$2\frac{1}{2}''$

0.400

1''

Impact value calculated on basis of 1-inch specimen, i.e., 1 inch notch length

$\frac{1}{2}''$

Specimen thickness (optional)

(G) Pointer and pendulum

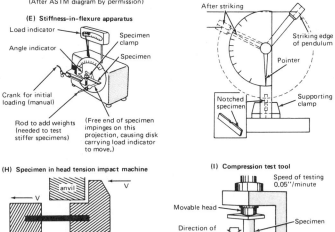

After striking

Striking edge of pendulum

Pointer

Notched specimen

Supporting clamp

(H) Specimen in head tension impact machine

anvil

V

V

Pendulum head

anvil

Crosshead clamp

(I) Compression test tool

Speed of testing 0.05''/minute

Movable head

Direction of load application

Specimen

Fixed head

Fig. 7-1. Test methods for plastics.

better for plastics—the tensile impact test. In the case of the Izod test, what is measured is the energy required to break a test specimen transversely struck (the test can be done either with the specimen notched or unnotched). The tensile impact test is done on a bar which is impact loaded in tension and the striking force tends to elongate the bar.

Plastics tend to be very notch sensitive on impact. This is apparent from the molecular structure of the materials which consist of random arrangements of polymer chains. If the material exists in the glassy state at room temperature the notch effect is to cut the chains locally and increase the stress on the adjacent molecular chains which will scission and propagate the effect through the material. At the high loading rate encountered in impact loading the only form of molecular response is the chain bending (spring) reaction which is limited in extent and generally low in magnitude compared to the viscoelastic response which responds at longer loading times.

There are several ways in which the impact properties of plastics can be improved if the material selected does not have sufficient impact strength. One method is by altering the composition of the material so that it is no longer a glassy polymer at the operating temperature of the part. In the case of vinyl chloride polymers this is done by the addition of an impact modifier which can be a compatible resin such as an acrylic or a nitrile rubber. The addition of such a material lowers the glass transition temperature and the material becomes a rubbery viscoelastic polymer with much improved impact properties. This is the way in which type II PVC materials are made to exhibit superior impact properties. Another way in which to improve impact properties is by orienting the material. Nylon has a fair impact strength but oriented nylon has a very high transverse impact strength. The intrinsic impact strength of the nylon comes from the polar structure of the material and the fact that the polymer is crystalline. The substantial increase in impact strength as a result of the orientation results from the polymer chains being aligned. This makes them very difficult to break and, in addition, the alignment improves the polar interaction between the chains so that even when there is a chain break the adjacent chains hold the broken chain and resist parting of the structure. The crystalline nature of the nylon material also means that there is a larger stress capability at rapid loading since the crystalline areas react much more elastically than the amorphous glassy materials.

Another way in which impact strength can be substantially improved is by the use of fibrous fillers. These materials, when bonded to the resin phase in the plastics material, act as a stress transfer agent around the region which is highly stressed by the impact load. Since most of the fibrous fillers such as glass and asbestos have high elastic moduli, they are capable of responding elastically at the high loading rates encountered in impact loading.

One general method of improving the performance of plastics parts in impact loading is to prevent, by design and handling, the formation of notched areas which act as stress risers. Especially under impact conditions the possibility of localized stress intensification can lead to part failure. Table 7-1 shows the notched IZOD impact strength of several plastics materials. In almost every case the notched strength is substantially less than the unnotched strength.

One last point about impact strength is its sensitivity to temperature. The impact strength of plastics is reduced drastically at low temperatures with the exception of fibrous filled materials which improve in impact strength at low temperature. The reduction in impact strength is especially severe if the material undergoes a glass transition where the reduction in impact strength is usually an order of magnitude.

A related form of stress to impact is impulse loading which differs in two ways from impact loading. Impact loading implies striking the object and consequently there is a severe surface stress condition present before the stress is transferred to the bulk of the material. In addition, in impact loading the load is applied instantly, limited in straining rate only by the elastic constants of the material being struck. A significant portion of the energy of impact is converted to heat at the point of impact and complicates any analytically exact treatment of the mechanics of impact. In the case of impulse loading the load is applied at very high rates of speed limited by the member applying the load. However, the loading is not generally localized and the heat effects are similar to dynamic loading in that the hysteresis characteristics of the material determines the extent of heating and the effects can be analyzed with reasonable accuracy. For example, the load of two billiard balls striking is definitely an impact condition. The load applied to a brake shoe when the brake is applied or the load applied to a fishing line when a strike is made is an impulse load. The time constants are short but not as short as the impact load and the entire structural element is subjected to the stress.

Table 7-1. Impact Resistance (Izod) of Plastics.

Polycarbonate	12.0 –16.0 ft lb/in
Polyester molding compound, glass-filled	1.5 –16.0
Vinyl, rigid, Type II	3.0 –15.0
Alkyd, glass-filled	8.0 –12.0
Cellulose propionate	0.8 –11.0
Polystyrene, impact	0.4 –11.0
ABS polymers, high-impact	5.0 –10.0
Polyethylene, high-density	1.0 –10.0
Phenolic molding compound, cord-filled	4.0 – 8.0
Cellulose nitrate	5.0 – 7.0
Ethyl cellulose	3.5 – 7.0
Epoxy molding compound, glass-filled	0.4 – 6.2
Polypropylene	0.6 – 6.0
Silicone molding compound, glass-filled	0.3 – 6.0
Diallyl phthalate, glass filled	6.0
PTFE-fluorocarbon	6.0
ABS polymers, medium-impact	0.7 – 5.0
Phenolic molding compounds, cotton fabric-filled	0.7 – 4.4
Cellulose acetate butyrate, Type I	1.1 – 4.3
Cellulose acetate, Type I	1.1 – 4.0
Polyvinylidene fluoride	3.8
PCTFE-fluorocarbon	3.6
Nylon 6	1.0 – 3.6
Cellulose acetate, Type II	0.4 – 2.7
Cellulose acetate butyrate, Type II	0.6 – 2.4
Acetal resins	1.0 – 2.3
Acrylic molding compound, modified	0.5 – 2.0
Diallyl phthalate, acrylic fiber-filled	1.2
Phenolic molding compound (rubber modified), cotton-filled	0.4 – 1.0
Vinyl, rigid, Type I	0.4 – 1.0
Polyvinylidene chloride	0.3 – 1.0
Melamine, fabric-filled	0.5 – 0.9
Nylon 66	0.9
Phenolic molding compound (rubber modified), wood flour-filled	0.34– 0.7
Polystyrene, heat and chemical resistant	0.3 – 0.6
Polystyrene, general-purpose	0.25– 0.6
Acrylic molding compounds	0.3 – 0.5
Diallyl phthalate, asbestos-filled	0.45
Melamine, mineral-filled	0.3 – 0.4
Chlorinated polyether	0.4
Phenolic molding compound (rubber modified), asbestos-filled	0.4
Phenolic molding compounds, mineral-filled	0.3 – 0.4

Table 7-1. (*Continued*)

Alkyd, mineral-filled	0.25– 0.35
Epoxy molding compound, mineral-filled	0.23– 0.35
Phenolic molding compounds, wood flour-filled	0.24– 0.35
Silicone molding compound, mineral-filled	0.25– 0.30
Styrene-methyl methacrylate copolymer	0.30
Melamine, cellulose-filled	0.24– 0.30
Urea, cellulose-filled	0.24– 0.30

Courtesy, E. I. du Pont de Nemours & Co., Inc., Wilmington, Del.

Plastics generally behave in a much different manner under impulse loading than they do under loading at normal straining rates. Some of the same conditions occur as under impact loading where the primary response to load is an elastic one because there is not sufficient time for the viscoelastic elements to operate. The primary structural response in the polymer is by chain bending and by stressing of the crystalline areas of crystalline polymers. The response to loading is almost completely elastic for most materials, particularly when the time of loading is of the order of milliseconds. Since all of the load is applied to the elastic elements in the structure and the long range strain adaptation is precluded, the material will exhibit a high elastic modulus and much lower strain to rupture. It is difficult to generalize as to whether the material is stronger under impulse loading than under normal loading. For example, polymethyl methacrylate and rigid polyvinyl chloride materials which appear to be brittle under normal loading conditions, exhibit high strength under impulse loading conditions. Rubbery materials such as thermoplastic urethane elastomers and some other elastomers behave like brittle materials under impulse loading. This is an apparently unexpected result which upon analysis is obvious because the elastomeric rubbery response is a long time constant response and the rigid connecting polymer segments which are brittle are the ones that respond at high loading rates.

The comments made with respect to impact loading for structures apply equally well to impulse loading conditions. Fibrous fillers improve impulse loading strength. Oriented materials withstand impulse loading much better than unoriented materials. Fibrous forms of materials are used in rope because they take impulse loading well. Crystalline polymers generally perform well under impulse loading, especially polar materials with high interchain coupling.

Using plastics under impulse loading conditions requires a careful design approach. Test data taken with high-speed testing machines are essential before using a polymer plastics material for these applications since it is difficult to predict the response of the material from the available data. Figure 7-2 shows a high-speed testing machine which is used to determine the response of materials at millisecond loading rates. In the absence of such test data, the only first sorting evaluation that can be done is from the results of the tensile impact test. The test should be done with a series of loads below break load, through the break load, and then estimating the energy of impact under the non-break conditions as well as the tensile impact break energy. As indicated above, apparently brittle materials perform

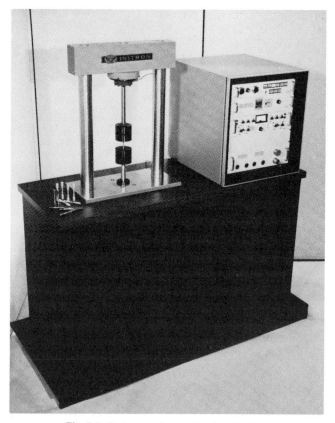

Fig. 7-2. Instron testing devices for plastics.

well and rubbery materials which would seem to be a natural for impulse loading behave in a brittle manner.*

The next category of stress that is of frequent interest is resistance to puncture. This type of loading is of particular interest in applications involving sheet and film as well as thin-walled tubing and other membrane type structures. The surface skins of sandwich panels are another area where it is important. Figure 7-3 illustrates what is meant by puncture stress. A localized force is applied by a relatively sharp object perpendicular to the plane of the sheet of material being stressed. If the material is thick compared to the area of application of the stress, it is effectively a localized compression stress with some shear effects as the material is deformed below the surface of the sheet. In the case of a thin sheet or film the stresses cause the material to be displaced completely away from the plane of the sheet and the restraint is by tensile stress in the sheet and by hoop stress around the puncturing member. Most cases fall somewhere between these extremes, but the most important conditions in practice involve the second condition to a larger degree than the first condition.

To analyze the second condition we take the material at the point where the puncturing object has almost pierced the membrane but has not broken through (Fig. 7-4). At this point we can see the nature of the forces which are resisting the puncture and qualitatively relate

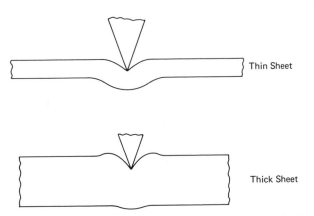

Thin Sheet

Thick Sheet

Fig. 7-3. Deflections by point puncture loading of thick and thin sheets.

*Reference *Modern Plastics* May 1964 article on repeated tensile impact.

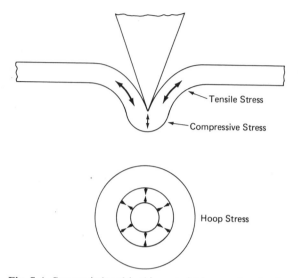

Fig. 7-4. Stresses induced in thin material by puncture loading.

them to the primary physical characteristics of the material so that we can indicate which materials are suitable for resistance to this type of stress and how we might improve the resistance to puncture. There are three principal stresses that result from the puncture forces through relatively thin material. There is a compressive stress under the point of the puncturing member, a tensile stress caused by the stretching of the material under the penetrating force, and, finally, there is a hoop stress caused by the material being displaced around the penetrating member: part of the hoop stress is compressive adjacent to the point which changes to tensile stress to contain the displacing forces. It is evident that anisotropic materials will have a more complicated force pattern and, in fact, uniaxially oriented materials will split rather than puncture under this type of loading. To improve the puncture resistance we need materials with high tensile strength. This is evident as required to have both the stretching load and the hoop stress. In addition, the material should have a high compression modulus to resist the point penetration into the material. Resistance to notch loading is also important.

Based on this analysis it is evident that materials which are biaxially oriented will have good puncture resistance. Highly polar polymers would be resistant to puncture failure because of their tendency to

Table 7-2.

Oriented nylon	2
Vulcanized fiber	6
Polyethyleneterephthalate	8
Cycolac DFA	10
Nylon	13
Mobay XEP-62	14
Surlyn 155	17
Estane	22

The above figures record the relative penetration of a sharp needle under static load into a variety of plastics. The lowest number indicates maximum puncture resistance.

increase in strength when stretched. The addition of randomly dispersed fibrous filler will also add resistance to puncture loads. Some materials which are notably resistant to puncture are listed in Table 7-2. From some examples such as oriented polyethylene glycol terephthallate (Mylar), vulcanized fiber, and oriented nylon, it is evident that these materials meet one or more of the conditions outlined above. Parts and materials which meet with puncture loading conditions in applications can be reinforced against this type of stress by use of a surface layer of material with good puncture resistance. Resistance of the surface layer to puncture will protect the part from puncture loads. An example of this type of application is the addition of an oriented polystyrene layer to foam cups to improve their performance.

The frictional properties of plastics are of particular importance to applications in machine parts and in sliding applications such as belting and structural units such as sliding doors. The range of friction properties is shown in the table of coefficient of friction values (Table 7-3). Figure 7-5 shows the relationship between the normal force and the friction force which is used to define the coefficient of static friction.

Friction coefficients will vary for a particular material from the value just as motion starts to the value it attains in motion. The coefficient depends on the surface of the material, whether rough or smooth, as well as the composition of the material. Frequently the molded surface of a particular plastics will exhibit significantly different friction characteristics from that of a cut surface of the same smoothness. These variations and others which are discussed below make it necessary to do careful testing for an application which relies

Table 7-3. Coefficients of Friction for Some Common Plastics.

Polymer	Steel on Polymer		Polymer on Polymer	
	μ_s	μ_k	μ_s	μ_k
PTFE ("Teflon") (polytetrafluoroethylene)	0.10	0.05	0.04	0.04
PTFE-HFP copolymer (FEP "Teflon") (tetra-fluoroethylene-hexafluoropropylene)	0.25	0.18	—	—
Polyethylene (low density)	0.27	0.26	0.33	0.33
Polyethylene (high density)	0.18	0.08–0.12	0.12	0.11
Acetal resin ("Delrin")	0.14	0.13	—	—
Polyvinylidene fluoride	0.33	0.25	—	—
Polycarbonate	0.60	0.53	—	—
PET ("Mylar") (polyethylene terephthalate)	0.29	0.28	0.27*	0.20*
Nylon (polyhexamethylene adipamide—	0.37	0.34	0.42*	0.35*
PFCE ("Kel-F") (polytrifluorochloroethylene)	0.45*	0.33*	0.43*	0.32*
PVC (polyvinyl chloride)	0.45*	0.40*	0.50*	0.40*
PVDC (polyvinylidene chloride)	0.68*	0.45*	0.90*	0.52*

*"stick-slip" (intermittent motion).

on the friction characteristics of plastics. Once the friction character-istics are defined, however, they are stable for a particular material fabricated in a stated manner.

The molecular level characteristics which create friction forces are the intermolecular attraction forces of adhesion. If the two ma-terials that make up the sliding surfaces in contact have a high degree of attraction for each other, the coefficient of friction is generally high. This effect is modified by surface conditions and the mechanical

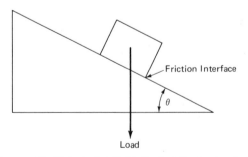

Fig. 7-5. Friction diagram. Coefficient of static friction is defined as $\tan \theta$ at the angle when the load starts to slide.

properties of the materials. If the material is rough there is a mechanical locking interaction that adds to the friction effect. Sliding under these conditions actually breaks off material and the shear strength of the material is an important factor in the friction properties. If the surface is not rough, but smoothly polished, the governing factor induced by the surface conditions is the amount of area in contact between the surfaces. In a condition of large area contact and good adhesion, the coefficient of friction is high. In the case of smoothly polished surfaces and adhesion forces the coefficient is very high since there is intimate surface contact.

Several other factors affect the frictional forces. If one or both of the contacting surfaces have a relatively low compression modulus it is possible to make intimate contact between the surfaces which will lead to high friction forces in the case of materials having good adhesion. It can add to the friction forces in another way as can be seen from Fig. 7-6. The displacement of material in front of the moving object adds a mechanical element to the friction forces.

All sliding friction forces are dramatically affected by surface contamination. If the surface is covered with a material that prevents the adhesive forces from acting, the coefficient is reduced. If the material is a liquid which has low shear viscosity we have the condition of lubricated sliding where the characteristics of the liquid control the friction rather than the surface friction characteristics of the

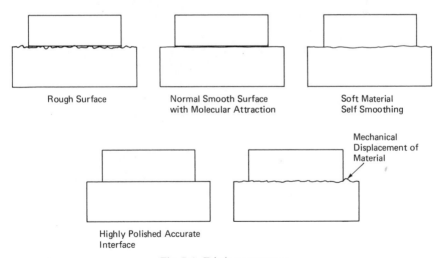

Fig. 7-6. Friction generators.

materials. It is possible by the addition of surface materials which have high adhesion to increase the coefficient of friction.

The use of plastics for gears (Fig. 7-7) and bearings is the area in which friction characteristics have been examined most carefully. Highly polar polymers such as nylon and the thermoplastic polyesters have, as a result of the surface forces on the material, relatively low adhesion for themselves and such sliding surfaces as steel. Laminated plastics (Fig. 7-8) make excellent bearings. The physical properties of these materials make them a good choice for both bearing and gear materials. The typical coefficient of friction for such materials is 0.1 to 0.2. In the molded condition the skin formed when the polymer cools against the mold tends to be harder and slicker than a cut surface so that the molded parts exhibit lower sliding friction and are excellent for this type of application. Good design for this type of application is to make the surfaces as smooth as possible without making them glass smooth which tends to increase the intimacy of contact and to increase the friction above that of a fine surface. The

Fig. 7-7. Plastics gear.

Fig. 7-8. Stern tube bearings for ships are made from phenalic canvas laminates. Similar bearings are used in water pumps and steel mill roll necks.

problems in this type of application related to friction are heat effects due to the rubbing surfaces. For successful design the heat generated by the friction must be dissipated as fast as it is generated to avoid overheating and failure.

Obviously the addition of appropriate lubricants will lower the friction and help to remove the heat. There are several other ways in which the friction can be reduced. One is by the incorporation of fillers. The fillers can be used to increase the thermal conductivity of the material such as glass and metal fibers. The filler can be a material like TFE polymer which has a much lower coefficient of friction and the surface exposed material will reduce the friction. Still another approach which is now used is the incorporation of slightly incompatible materials such as silicone oil into the molding material. After molding the material migrates to the surface of the part and acts as a renewable source of lubricant for the part. In the case of bearings it is carried still further by making the bearing material porous and filling it with a lubricating material in a manner similar to sintered metal bearings, graphite and molybdenum sulfide are also incorporated as solid lubricants.

A different type of low friction or low drag application is encountered with sliding doors or conveyor belts sliding on support surfaces. In applications like this the normal forces are generally quite small and the friction load problems are of the sticking variety. Some plastic materials exhibit excellent track surfaces for this type of application. TFE resins have the lowest coefficient of any solid material and represent one of the most slippery surfaces known. The major problem with TFE is that its abrasion resistance is low so that most of the applications utilize filled compositions with ceramic filler materials to improve the abrasion resistance. There is a whole field of applications for TFE materials in reducing friction using solid materials as well as films and coatings. Another material with excellent properties for surface sliding is ultra high molecular weight polyethylene. Polyethylene and the polyolefins in general have low surface friction, especially against metallic surfaces. The UHMW material has an added advantage in that it has much better abrasion resistance and is preferred for conveyor applications and applications involving materials sliding over the part. In the textile industry loom parts also use this material extensively because it can handle the effects of the thread and fiber passing over the surface with low friction and relatively low wear.

The specific friction characteristics of plastics at the high friction end is also an area of significant applications. Some plastics, notably polyurethane and some plasticized vinyl compositions, have very high friction coefficients. These materials make excellent traction surfaces for parts ranging from power belts to drive rollers where the plastics part either drives, or is driven, by another member. Conveyor belts made of oriented nylon and woven fabrics are coated with polyurethane elastomer compounds to supply both the driving traction and to move the objects being conveyed up fairly steep inclines because of the high friction generated. Drive rollers for moving paper through printing presses and business machines are frequently covered with either urethane or vinyl to act as the driver members with minimum slippage. The materials are also used as the torque surfaces in clutches and brakes.

In all of the friction applications suggested as well as in others, there are two areas where the design effort is introduced. The first is in material selection and modification to provide either high or low friction as required by the application. The other is in determining the required geometry to supply the frictional force level needed by controlling contact area and surface quality to provide friction level. A controlling factor limiting any particular friction force application is heat dissipation. This is true if the application of the friction loads is either a continuous process or a repetitive process with a high duty cycle. The use of cooling structures either incorporated into the parts or by the use of external cooling devices such as coolants or air flows should be a design consideration.

Another form of stress, again related to the impulse and impact loading, is the effect of erosion forces such as wind driven sand or water, underwater flows of solids past plastic surfaces and even the effects of high velocity flows causing cavitation effects on material surfaces. One major area for the utilization of plastics is on the outside of moving objects that range from the front of automobiles to boats, aircraft, missiles, and subsea craft. In each case the impact effects of the velocity driven particulate matter can cause surface damage. Stationary objects such as radomes and buildings exposed to the weather in regions with high and frequent winds are also exposed to this type of effect.

Figures 7-9, 7-10 and 7-11 show one type of wind erosion analysis which has been extensively studied. This is the effect of water drop erosion on rapidly moving missile parts. Aircraft radomes have also been extensively studied for the effects of wind-driven water and solids. The erosion effects are very dramatic and the surfaces are usually protected with materials that have good resistance to this type of stress.

To determine the type of physical properties materials used in this environment should have, it would be necessary to examine the mechanics of the impact of the particulate matter on the surfaces. Figure 7-12 shows what takes place when a water drop impacts a surface. The high kinetic energy of the droplet is dissipated by shattering the drop, by indenting the surface, and by frictional heating effects. The loading rate is high as in impact and impulse loading, but it is neither as localized as the impact load nor as generalized as the impulse load. Material which can dissipate the locally high stresses through the bulk of the material will respond well under this type of load. The material should not exhibit brittle behavior at high loading rates.

Fig. 7-9. The most severe loading experienced by an artillery fuze radome is high velocity flight through a tropical rainstorm. Some materials which have shown acceptable resistance to erosion at modern artillery muzzle velocities are modified polyphenylene oxide, polyimide, polysulfone, and irradiated polyethylene. Protective caps must be provided for muzzle velocities above 2700 fps because surface temperatures approach 700° F and droplets easily penetrate the ¼ in. thick tip of the radome and destroy the fuze electronics within. Theories in the literature describing stresses generated by raindrop impact vary widely when magnitudes are calculated, so experimental approaches are used exclusively in design. One method is to use the rocket sled—artificial rainfield facility at Holloman Air Force Base. The length of track and booster system is sufficient to provide the proper aerodynamic heating of the radome prior to entering the rainfield test area, which in turn has a length and nozzle system that can reasonably simulate a tropical storm. (*Courtesy Harry Diamond Laboratories*)

In addition, it should exhibit a fairly high hysteresis level which would have the effect of dissipating the sharp mechanical impulse loads as heat. The material will heat due to the stress under cyclical load.

There is another characteristic to be considered in this type of loading. The surface properties of the material are quite significant. If the water does not wet the surface, the tendency would be to have the droplets that do not impact close to perpendicular bounce off the surface with considerably less energy transfer to the surface. Non-wetting coatings reduce the effect of wind and rain erosion.

Impact of air-carried solid particulate matter is more closely analogous to straight impact loading since the particles do not flow or become disrupted by the impact. The main characteristic required of the material, in addition to not becoming brittle under high rate load-

Fig. 7-10. Radomes for artillery fuzes are made from thermoplastic materials because the cost is low and the physical properties are electrically and mechanically acceptable for modern weapons. Typical thermo-mechanical loadings are 20,000g setback during gunfire, 300 rps spin and 3000 fps muzzle velocity, 700° F peak surface temperature during flight, and a range of −40° F to +160° F storage temperatures. Materials such as modified polyphenylene oxide, polysulfone, polyimide, polycarbonate, (i.e. materials with high heat of distortion temperatures) are currently in use. Two factors will limit the application of thermoplastics to futuristic, extended range guns, however: 1) Mechanical ablation (as contrasted to thermal ablation) of material due to high surface temperatures, windstream shear, and spin; and 2) Rain erosion resistance. The picture shows glass fiber filled modified polyphenylene oxide radomes for the artillery fuze after wind tunnel tests simulating gunfire at progressively higher muzzle velocities. Tip failure occurs first, material stripping along the sides is second, and after 20 seconds exposure to 3500 fps wind, the radome extruded from the metal crimp joint. Design concepts now are geared to protective caps on the tip of the radome rather than turning to the more expensive thermosetting plastics. One major design problem is that there is no theory for mechanical ablation in the literature. Most attention is devoted to re-entry vehicles which experience thermal ablation due to the very high velocities. (*Courtesy Harry Diamond Laboratories*)

ing, is resistance to notch fracture. The ability to absorb energy by hysteresis effects is also important as is the case with the water. In many cases the best type of surface is an elastomer with good damping properties and good surface abrasion resistance. Polyurethane coatings and parts are excellent for both water and particulate matter that is air-driven. Besides such applications as vehicles, these materials are used in the interior of sand and shot blast cabinets where they are constantly exposed to this type of stress. These materials are fabricated into liners in hoses for carrying pneumatically conveyed materials such as sand blasting hoses and for conveyor hose for a wide variety of materials such as sand, grain, and plastics pellets.

The cavitation effects that occur when liquids such as water impact the surfaces of submerged vehicle parts induce a somewhat different type of stress on the materials of the structure. The loading is impulse

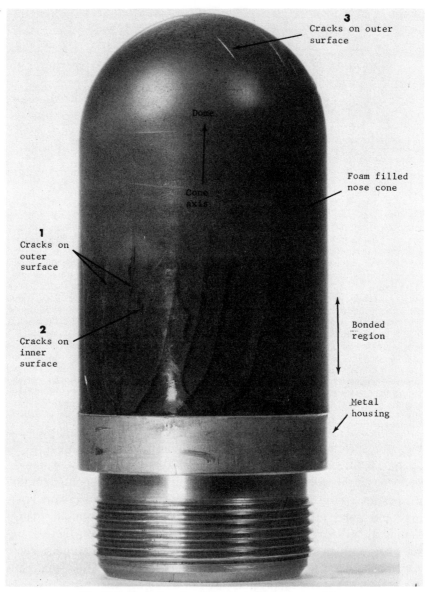

Fig. 7-11. Typical crack pattern on a nose cone assembly.

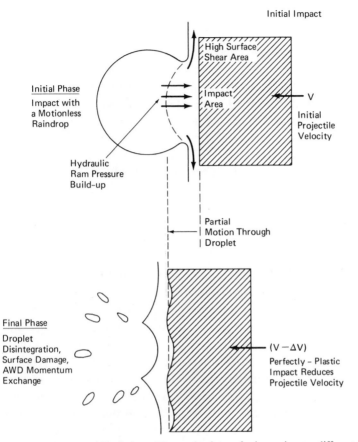

Fig. 7-12. Projectile impact with raindrop. The mechanisms of rain erosion are different from high and low velocity flight. Stress waves are important at high velocities; this must be analyzed using the projectile as the moving body rather than the raindrop. The impact is in the order of milliseconds. For further study see George K. Lucey Jr. A rain impact analysis for an artillery system. (*Courtesy Harry Diamond Laboratories*)

in nature but the stress can be negative as well as positive. The fluid passing the surface can cause negative pressure conditions which tend to lift the surface of the material alternating with impact loads as the vacuoles produced by cavitation collapse and induce compressive surface impulse loading. The alternating stress condition is more severe than the impingement process and can lead to spalling of material from the surface of the structure. Energy absorbing structures with good resilience that do not exhibit brittle behavior at impulse loading rates are again preferred for this type of structure. Elastomeric plastics make excellent parts and part coatings for resistance to cavitation loading.

In general. when the surface impact loading by fluid-borne particulate matter, liquid or solid, or cavitation loading is encountered, the method of minimizing the effects of erosion produced are by material selection and modification. The materials used should be ductile at impulse loading rates and capable of absorbing the impulse energy and dissipating it as heat by hysteresis effects. The surface characteristics of the materials in terms of wettability by the fluid and frictional interaction with the solids also play a role. In this type of application the general data available for materials should be supplemented by that obtained under simulated use conditions since the properties needed to perform are not readily predictable from the usually available data.

One other type of loading condition that will be discussed briefly because of its increasing interest in subsea applications is the application of external hydrostatic stress to plastic structures. Internal pressure applications such as those encountered in pipe and tubing or in pressure vessels such as aerosol containers are easily treated using tensile stress and creep properties of the materials with the appropriate relationships for hoop and membrane stresses which are summarized in *Formulas for Stress and Strain.** The application of external pressure, especially high static pressure, has a rather unique effect on plastics materials. The stress analysis for thick walled spherical and tubular structures under external pressure are also summarized in Roarke. The interesting aspect that plastics have in this situation is that the high compressive stresses increase the resistance of polymer-based materials to failure. Glassy polymers under conditions of very high hydrostatic stress behave in some ways like a compressible fluid. The density of the material increases and the compressive strength is increased. In addition, the material undergoes sufficient internal flow to distribute the stresses uniformly throughout the part and, as a consequence, the plastics parts produced from such materials as methyl methacrylate polymer make excellent view windows for undersea vehicles that operate at extreme depths where the

*Roarke, McGraw-Hill, 1965.

external pressures are 1000 psi and more. Other applications of this unique pressure stiffening characteristic will undoubtedly be found.

This chapter has covered some of the stress loading situations other than simple tensile, compressive and bending loads encountered by plastics parts. The unique characteristics of plastic materials of certain types make them especially suited to resist the specific stress conditions. With proper part design, material selection and modification the designer can make parts that will perform well under unique stress conditions.

Chapter 8

Design for Stiffness

In structural applications for plastics, which generally include those in which the part has to resist substantial static and dynamic loads, one of the problem design areas is the low modulus of elasticity of polymeric materials. Even when such rigid polymers as the ladder types of polyesters and polyamides are considered, the elastic moduli of unfilled polymers are under one million psi as compared to metals where the range is usually 10 to 40 million psi. Ceramic materials also have high moduli. Since shape integrity under load is a major consideration for structural parts, plastics parts must be designed for efficient use of material to afford maximum stiffness.

In the previous chapters we covered the use of material modification such as orientation and the use of fillers* to increase the modulus of elasticity of plastics. This chapter is concerned with geometrical design which makes the best use of materials to improve stiffness. Structural shapes which are applicable to all materials are discussed such as sandwich structures, shells, and folded plate structures. In the case of plastics, emphasis is on the way plastics can be used in these structures and why they are preferred over other materials. In many cases plastics can lend themselves to a particular field of application only in the form of a sophisticated lightweight stiff structure and the requirements are such that the structure must be of plastics, e.g., in a radome. In other instances, the economics of fabrication and erection of a plastics lightweight structure and the intrinsic appearance and other desirable properties make it preferable to other materials.**

One of the most widely used lightweight structural concepts is the

---

*Lees, J. K., A Study of the Tensile Strength of Short Fiber Reinforced Plastics. *Polymer Science Engineering*, **8** (3) July 1968.

**Lubin, George, *Handbook of Fiberglass and Advanced Composites*, 1969. John Milewski comment.

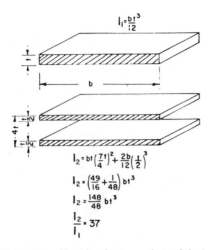

Fig. 8-1. Comparison of moments of inertia of cross sections of single slab and double slab of half thickness. The moment of inertia has been computed with reference to the central horizontal plane, that is, "neutral axis" in each case.

sandwich panel. From Fig. 8-1 it can be seen that the basic concept is to use two thin layers of material (Figs. 8-2 and 8-3) which are held separated by a very low density spacer of some kind. Figure 8-1, which does not show the nature of the separator, gives the relative stiffness in bending of a solid sheet of material and the same thickness of material made into a sandwich panel. It is well to note that the improvement in performance of a sandwich panel is limited essentially to its resistance to bending. There are some other criteria that the panel must meet in order to have structural integrity. The skins and core must co-act in order for the improvement in stiffness to hold and, consequently, the shear strength of the interface between the core and skin must be sufficiently high to prevent shear failure. The shear strength of the core material must be sufficient to resist the shear stresses imposed by bending so that it does not fail at stress levels below that at which the skin materials fail. The skin-to-core bond must also be capable of sufficient stress to handle the buckling forces imposed on the skin by the bending of the panel.

The above are requirements for a sandwich panel to perform its improved function in bending in the direction perpendicular to the plane of the panel. A sandwich panel exhibits no improvement in performance in other directions such as parallel to the plane of the sandwich. It is, in fact, subject to failure under lower load conditions

Fig. 8-2. This process permits geometric configurations from a variety of modern plastics for optimum structural properties as well as unique decorative effects. The novel features are applicable to new designs in constructions, containerization, partition systems, energy absorption barriers, furniture, and a variety of other applications demanding light-weight structural strength and decorative innovation. (*Courtesy Northfield Corp.*)

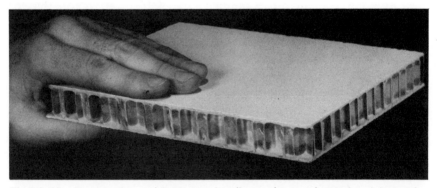

Fig. 8-3. New honeycomb material enters truck trailer, marine container and modular housing field. Polyester-glass reinforced sandwich cuts fabrication costs, maintenance, and reduces weight by more than 20%. A new paper honeycomb sandwich material beefed up by fiber glass reinforced polyester resin is now making inroads into three markets presently dominated by aluminum, steel, wood and other materials. Pictured above is a view of the new material which essentially consists of fiber glass reinforced polyester skins around a phenolic-impregnated paper honeycomb interior. (*Courtesy A. J. Lazarus Associates, Inc.*)

under edge loading because of the susceptibility to failure of the skins by buckling. Despite these limitations structural sandwich panels are a very efficient lightweight structural element widely used in buildings, in aircraft, in surface vehicles, and in many commercial and industrial

Fig. 8-4. Structural foam molded parts are molded with skin surfaces and foamed core. (*Courtesy General Electric Co.*)

applications. Figure 8-2 shows a number of different core materials which are used with sandwich panels such as foam, honeycomb core, ribs, filler spacers, corrugated sheet spacers, etc. Plastics, such as glass-reinforced polyester, are frequently used as the skins for panels. Plastics materials such as polyurethane foam, cellulosic foams, and others are widely used as core materials. In some cases metallic skins are used in conjunction with plastics or plastic foam cores. The combinations, from identical core and skin materials using expanded material for the core and solid material for the skin-to-metal sheathed plastic parts made by plating (using the plastic core and metallic skins) are the extremes with a wide range of possibilities in between.

Figures 8-3, 8-4 and 8-5 show some of the parts which incorporate the sandwich panel concept to improve stiffness. They range from structural foam molded parts (which come from the mold as completed parts) incorporating low density cores and high density skins of the same materials to parts vacuum formed of a plastics material, the core of which becomes cellular during the heating process for forming. Reinforced plastic structural panels for curtain wall building construction that have honeycomb cores or decorative cores are examples

Fig. 8-5. Workers installing a composite plastics roof system. (*Courtesy Uniroyal*)

of the widespread use of the panel concept. Plated plastics parts covered with substantial layers of metal are another use of reinforcing skins to improve the stiffness of plastics parts as well as to improve the appearance and environmental resistance. Extruded cellular plastics shapes with applications that range from molding substitutes for wood to structural shelving are other examples where the sandwich panel stiffening principle is applied.

Another widely used geometry employed to improve stiffness is the use of corrugated and dimple sheet surfaces. Figure 8-6 shows several examples, including corrugated sheet and a complex dimple pattern which can improve the flexural stiffness in one or more directions. In each case the improvement in flexural stiffness is made by displacing material from the neutral plane. This increases the EI product which is the geometry material index that determines resistance to flexure.

Another type of stiffener is shown in Figs. 8-7 and 8-8. In this case the basic sheet of the part is converted to a series of connected I or T beams. While this construction is not as efficient as the sandwich panel, it does have the advantage that it can be molded or extruded directly in the required configuration and the relative proportions of the legs and sheet can be designed to meet the flexural requirements. One of the limitations is that it imparts increased stiffness in one direction much more than in the other.

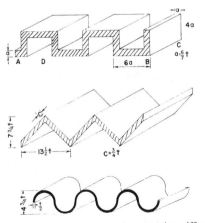

Fig. 8-6A. These corrugated sheets are equivalent in crosswise stiffness to ribbed section but are not nearly as stiff in lengthwise direction.

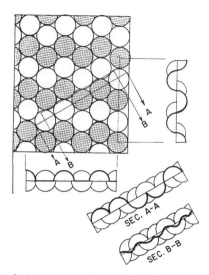

Fig. 8-6B. Staggered hemisphere pattern offers good stiffness in most directions. Heavy lines in sectional views indicate cross sections in planes of cuts. Section B-B has low moment of inertia, low stiffness, would fold relatively easily.

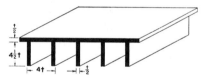

Fig. 8-7. T-beam construction used in most home roofing is also useful in stiffening the surfaces of plastics parts.

Fig. 8-8. Bottom flanges on T-beams convert them to I-beams, which are stiffer in crosswise direction.

Figures 8-9 and 8-10 show two forms of egg-crate design that overcome the limitations posed by the ribbing technique to make T and I beam sections. The stiffness improvement in this case is isotropic, essentially in the plane of the structure, but it is less than that of the I beam T beam configuration in the preferred direction when the same amount of material is used.

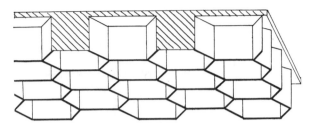

Fig. 8-9. Honeycomb ribbing attached to flat surface gives excellent resistance to bending in all directions.

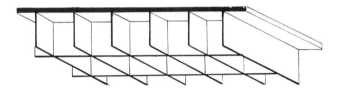

Fig. 8-10. Egg-crate ribbing also provides multidirectional stiffening, though not quite as isotropically as honeycomb.

Figures 8-11 and 8-12 show several novel ways that a structural sandwich can be introduced into the design of molded parts. In each case the part is molded of two pieces which are then joined by an adhesive or some form of welding such as ultrasonic welding. When

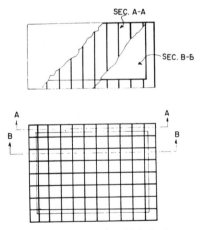

Fig. 8-11. A double-walled, box-like structure in which the bottom members are stiffened by egg-crate ribbing; the sides are stiffened by vertical ribs.

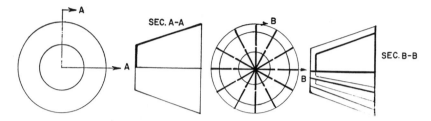

Fig. 8-12. Double-walled conical structure uses either foam or ribs to effect stiffening separation and linking of walls.

using these stiffening techniques it is important to realize that the same criteria with regard to bond between the skins and the spacer material must be met in any sandwich construction. If the bond fails, there is no structural integrity.

We will now examine a much broader group of shapes which impart rigidity to a structure. These are shapes which have intrinsic rigidity due to a specific and not a repetitive geometry. The two simplest ones are shown in Figs. 8-13 and 8-14. Figure 8-13 is a frustrated cone which is loaded on the smaller face. The conical section, by virtue of the tapered sides, stands up under large compressive loads without loss of the buckling stability which would occur with a cylinder used to replace it. The incorporation of a stiffener at the lower edge increases the resistance to the load.

Figure 8-14 shows the use of transverse shaped ribs to improve the stiffness of a plane surface loaded in compression on the top with the corners supported. The ribs are, in effect, flying buttresses which

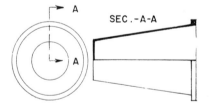

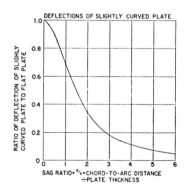

Fig. 8-14. Radial tapered rigs help stiffen this slab, also carry load to support points at corners.

Fig. 8-13. Conical container gains much lateral stiffness by simple bead around lip.

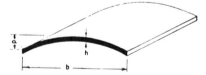

Fig. 8-15 A. Sketch of cylindrially curved plate.

transfer some of the local corner loads to the center part of the sheet and at the same time increase the resistance of the sheet to bending. While a rectangular sheet is shown, the method is applicable to polyhedral sheets with any number of sides as well as to circular elements. The elements do not need to be rotationally symmetrical and the elements can be rectangles or ellipses or any other suitable shape. The geodesic dome uses a set of these elements in conjunction with a space lattice to form strong structural units.

The concept of the geodesic dome brings us to a class of structures which rely for their load bearing ability on two and three dimensional forms. Figure 8-15A shows a curved plate which is loaded against the

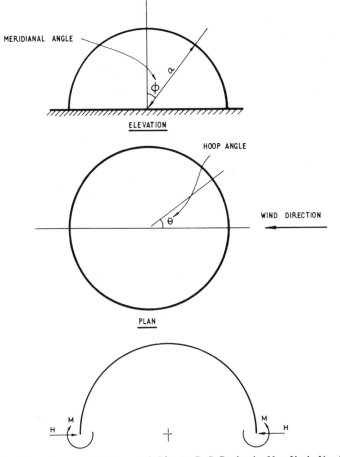

Fig. 8-15B. (From *Structural Designs with Plastics,* B. S. Benjamin. New York, Van Nostrand Reinhold, 1969.)

convex surface of the plate. Figure 8-15A shows how the resistance to deformation is related to the amount of curvature of the plate. This is shown for plates with slight curvature and indicates that some of the stress perpendicular to the plane of the plate is converted to stress in the sheet direction. It should be apparent that when the degree of curvature is large, we are dealing with an arch where the load is completely transferred from bending loads on the arch to compressive loads on the arch supports.

Domes are another group of geometric shapes that are used as structural elements to support distributed loads with minimum bending. The effect of the dome structure is to change the bending stresses caused by the loads into compressive stresses in the dome sheet. These stresses are transferred to the support points of the dome where they are then transferred to the supporting elements as shown in Figure 8-15B. The termination point of the spherical shell has some rather complex stresses imposed. This is the point in the structure that must be analyzed to determine limiting loads. In general, there are two limitations on the loading of shells of this type. One is that the localized stresses must be kept below the values which will cause puncture of the shell. The other limitation is generated by the buckling effect in the shell caused by the stresses transferred through the shell. An excellent analysis of shells made from plastics materials is given by Benjamin.*

Plastic skydomes which are formed from transparent plastic sheets are a good example of the use of this form. The resistance of this shape to external wind and snow loads is much greater than a structure made from a flat sheet or a combination of flat sheets of material. The economy of these units has made plastics a strong competitor of glass for use in skylights despite the relatively high cost of plastics compared to glass (Fig. 8-16).

The increasing use of plastics in large structures such as radomes, space structures, and architectural structures particularly, has resulted in a number of interesting and unique types of stiff structures with a somewhat complicated geometry. These designs are also used in structures made of other materials such as reinforced concrete and in structures made of combinations of plastics and other materials. Some of these shapes are shown in Figs. 8-17, 8-18, 8-19 and 8-20. One of

---

*B. S. Benjamin, *Structural Design with Plastics*, Van Nostrand Reinhold, New York, 1969.

Fig. 8-16. Spherical segment domes of acrylic plastics are used for the roof over a swimming pool. (*Courtesy Rohm & Haas Co., Philadelphia, Pa.*)

these is the hyperbolic paraboloid shown in Fig. 8-21. This structure uses tension types of structural elements to sustain the imposed distributed loads on the extensive surface of the unit. This is a particularly efficient use of materials since there are many materials such as steel wire, fiber glass cables, and oriented high polymers such as nylon which have very high tensile strength for their weight. In addition, the use of tensile strength loading eliminates the possibility of failure by buckling which is a problem with both sandwich structures and the two- and three-dimensional shell structures. Another related group of geometries is used in stressed skin structures, of which one of the classic examples is the aircraft wing shown in Fig. 8-22. In this structure the function of the framework is to maintain the skin membrane in

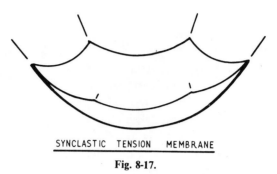

SYNCLASTIC   TENSION   MEMBRANE
Fig. 8-17.

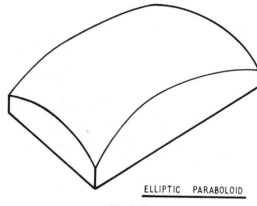

ELLIPTIC   PARABOLOID
Fig. 8-18.

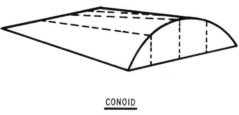

CONOID
Fig. 8-19.

Figs. 8-17 through 8-21 from *Structural Designs with Plastics*, B. S. Benjamin. New York, Van Nostrand Reinhold, 1969.

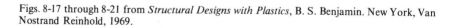

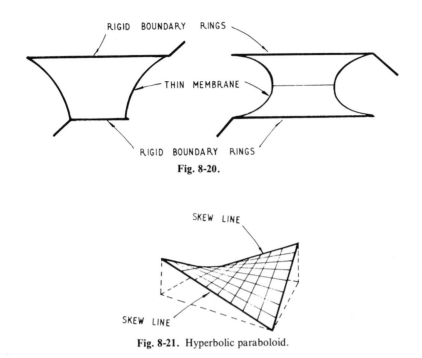

Fig. 8-20.

Fig. 8-21. Hyperbolic paraboloid.

shape so that it can take the necessary loads imposed. In the example of the wing, the loads imposed are aerodynamic as shown in Fig. 8-23 which give the wing its lift characteristic and the wing root bending loads where the lift forces are transferred to the main body of the aircraft. By prestressing the skin of the wing, all of the stress is in tension. When this condition is maintained, the possibility of failure by buckling is reduced or eliminated. The only region where careful analysis is required is at the root of the wing where a rigid transfer

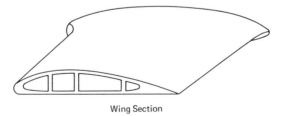

Wing Section

Fig. 8-22. Wing section.

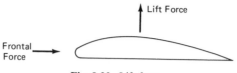

Fig. 8-23. Lift force.

structure is required. An airplane with an all plastics wing using the stressed skin has just gone into production. This indicates that it is a useful area for glass fiber or carbon filament reinforced plastics. By applying filament winding methods and using unidirectionally oriented reinforcement sheets, tensile strengths of 100,000 psi and moduli of 5×10^6 to 10^7 psi are achieved. The material density is in the range of 1.5 to 1.8, making these the highest strength-to-weight materials commercially available. Wider application of the concepts of stressed skin structures using biased reinforcements in the direction of the primary stresses will lead to new applications in vehicles and large lightweight structures such as hangers and cover domes for sports stadiums and similar recreation areas.

The use of tensile loading is important in many areas of the structures which employ plastics materials. Air-supported structures make use of many plastics materials. While the principles of air-supported structures apply to all types of applications, it is convenient from an end-use point of view to classify them into space structures— outer space, that is—and air-supported structures on the ground.

Space exploration has given a strong impetus to the design of lightweight air pressure erectable structures. The high energy cost to loft a unit of mass into orbit coupled with the low functioning stress level on orbiting structures makes any lightweight structure desirable. The gas pressure erection technique or a similar method is necessary since the packages must be compact to be stored in the launch vehicle. Gas pressure erection techniques are simple and very general in their applicability compared with mechanical or electronic techniques. They also have high reliability.

Figure 8-24 shows some designs of space erectable structures. In each case the gas pressure moves the materials that form the skins and ribs into final position. If the structural material is an unreacted resin plus reinforcement, the structure can be rigidized permanently by curing the resin in place as illustrated in Fig. 8-25. This is done by several methods such as the use of actinic light from the sun with

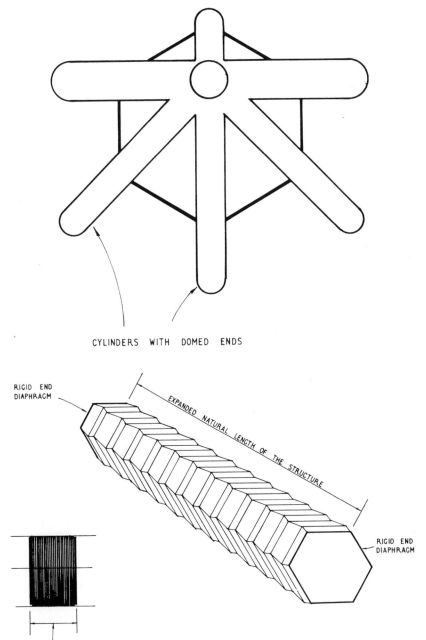

CYLINDERS WITH DOMED ENDS

RIGID END
DIAPHRAGM

EXPANDED NATURAL LENGTH OF THE STRUCTURE

RIGID END
DIAPHRAGM

PACKAGED LENGTH OF THE STRUCTURE

Fig. 8-24. Designs for space erectable structure. (From *Structural Designs with Plastics*, B. S. Benjamin, New York, Van Nostrand Reinhold, 1969.)

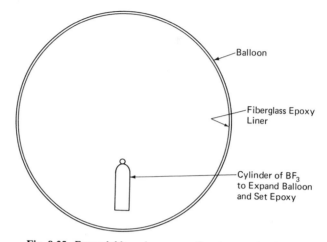

Fig. 8-25. Expandable and permanently set space structure.

appropriate catalyst in the resin, the application of gaseous catalysts contained in the gas that is used to expand the structure (e.g., BF_3 with epoxy resins), or the use of heat generalized either by embedded heating conductors in the structure or by radiant heating applied to the structure. As efforts continue to launch larger space laboratories, this type of design activity is destined to increase and, in the unusual environment of space, will undoubtedly lead to some interesting and new geometrical concepts for structures. Just as earth viewed from space creates a new perspective, the view of structural needs will generate new design perspectives.

The use of air-supported structures in general has resulted in many interesting applications, and are just beginning. The basic concepts that air-supported structures utilize are the same as those for tents, they are easily erected, easily dissembled, portable shelter units that can be used as houses, warehouses, temporary shelters for permanent construction or pool covers. At present air-supported structures are made from woven fabric webs coated with flexible polymer materials that range from polyethylene and EVA to elastomers of various types. The woven webs supply the oriented high-strength tensile members, and the coatings, the necessary gas barrier. In some instances, materials such as biaxially oriented polyester (i.e., Mylar) are used without reinforcement since the oriented film is intrinsically a high strength tensile member.

Fig. 8-26. An air supported warehouse. (*Courtesy Dynamic Displays*)

Figures 8-26 and 8-27 show several types of air-supported structures. Figure 8-26 shows a simple arch section with spherical sections ends. The perimeter of structures, frequently made of such materials as chlorosulfonated polyethylene (Hypalon elastomer) coated on nylon

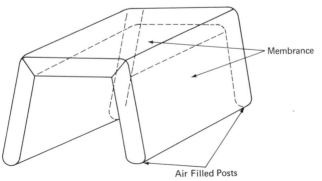

Fig. 8-27. Air-supported structure.

cloth, is a tubular element filled with water to hold the structure down. One or more blowers are used to fill the structure with air at a pressure in the range of 5 to 10 inches of water (about .2–.4 psi). The doors are of an air-lock type, either a revolving door or two sequential interlocked doors to keep the pressure in. The blowers can either operate continually or be controlled by a pressure switch. Despite the low internal pressure, the structure is stiff enough to support substantial roof loads, both static and dynamic. At .25 psi internal pressure the roof loading can be 36 psi which is adequate in most parts of the world for snow and wind loads. Obviously slightly higher pressures can be used to compensate for higher loads.

Figure 8-27 shows the next step in the evaluation of air-supported structures. It consists of gas prestressed beam members. The beams are connected by membranes to form either a single membrane or a double membrane structure. The gas-supported beam units are generally made from higher strength sheet materials and the internal pressure levels are in the 1 to 2 psi range. The membrane sheets are attached to the beam members so that in the erected structures they are under tension to form smooth walls capable of supporting substantial structural loads.

This design has several important advantages over the simple air-supported structure. The range of shelter shapes is increased substantially. Rectangular, round, polyhedral, and other shapes are possible. High ceilings are easy to achieve with little wasted space. The structure is more readily anchored since only the ends of the support beams must be tied down. The use of double membrane structures permits insulation against the external temperature changes. Since the structure is supported by the gas-filled structural beams or other elements, the need for air-lock doors is eliminated.

Air-supported structures of both types have been used as forms on which to cast concrete structures. In addition, some developmental work has been done to rigidize such enclosures by the use of reactive resin systems. In addition to the catalyzed setting up of the resin in the membrane, as in the space structure approach, urethane foam materials (and other reactive foams) have been introduced into the beam elements to replace the gas.

There are many variations possible with the gas-supported or gas-erected structure concept, many of which have not yet been tried. The basic concept is applicable to parasols, vehicle covers, awnings, and a host of other items.

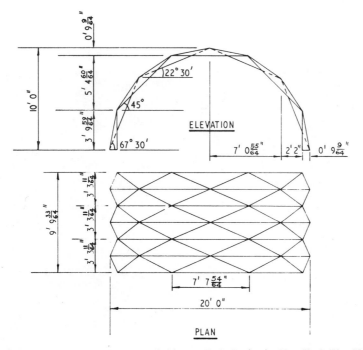

ELEVATION

PLAN

Fig. 8-28. (From *Structural Designs with Plastics*, B. S. Benjamin, New York, Van Nostrand Reinhold, 1969.)

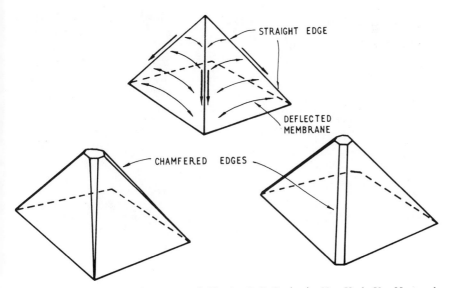

STRAIGHT EDGE

DEFLECTED MEMBRANE

CHAMFERED EDGES

Fig. 8-29. (From *Structural Designs with Plastics*, B. S. Benjamin. New York, Van Nostrand Reinhold, 1969.)

Fig. 8-30. FRP barrel vault unit undergoing test. (From *Structural Designs with Plastics*, B. S. Benjamin, New York, Van Nostrand Reinhold, 1969.)

Fig. 8-31. An amphibious cross-country vehicle of fiberglass sandwich construction. (*Courtesy Messerschmitt-Bolkow-Blohm GMBH (MBB)*)

The use of folded plates to form stiff structures has become one of the most important developments in architecture, as much because of the variety of interesting enclosures and support shapes possible as the efficient use of materials to impart stiffness. Benjamin gives a large number of examples of folded plate structures using FRP materials. They are one of the preferred materials for this type of construction because they can be readily fabricated into the required complex geometrics.

Figure 8-28 shows a basic barrel vault configuration which is one of the most widely used folded plate units. Figure 8-29 shows a dome construction of folded plates. Figure 8-30 shows a FRP barrel vault unit undergoing testing.

There are a number of other geometrics that yield structures with high stiffness using low modulus materials such as plastics. The examples given in this chapter include the more commonly used types. The important point to be made is that by the exercise of design ingeniuty, stiff lightweight structures can be designed and built

Fig. 8-32. A 68,380-square-foot tension structure project is being built at La Verne College in California. Roofs are of "Fiberglas" fabric coated with a new formulation which incorporates "Teflon" fluorocarbon resin. (*Courtesy E. I. Du Pont de Nemours & Co., Inc.*)

which offer economy, beauty, durability, and resistance to stress. Figure 8-31 shows a jeep body of self-supporting fiberglass. Plastics make excellent materials for these applications, especially if their directional properties can be utilized. While most of the present applications and examples are for architectural shapes (Fig. 8-32) there is good reason to apply the same approach to furniture, appliances (including refrigerator cabinets where the foam is also an insulator), vehicles, and other applications to insure efficient use of materials in prudent design.

Chapter 9

Processing Limitations on Plastics Product Design

Plastics parts are made by a variety of manufacturing processes. A detailed discussion of these is outside the scope of this book on plastics product design. However, it is essential to know some of the process-imposed limitations or an unmanufacturable product will be designed. In this chapter we will examine some of the general limitations imposed on design by the processes used to make the part.

Table 9-1 lists a number of methods used to manufacture plastic parts and the general types of shapes that are made by the process. The most important processes are molding methods that include injection, blow, extrusion, rotational, thermoforming, compression, and transfer molding. In each of these processes, as well as some of the others, the part is produced in a mold (an enclosure that is the boundary surface of the part), and the use of the mold introduces a series of limitations on the shape of the molded part.

A typical mold shape is shown in Fig. 9-1. The mold separates into two or more sections so that the part can be removed. In order to allow the part to be removed from the mold, certain geometrical considerations must be met. First, the part should have no undercut sections that will lock if the part is pulled from the mold. Figure 9-2 shows an example of an undercut. If the shape is essential to function, a much more complicated mold is required where a portion of the mold is retracted to permit the undercut to be removed. This complicates the molding procedure and the mold, and may result in higher costs as well as a poorer quality part. The part will usually have some surfaces that are nearly parallel and perpendicular to the opening surface of the mold (generally referred to as the parting line) and pulling the part against these parallel surfaces will result in sticking and drag that will make removal difficult and damage to the product

Table 9-1. Manufacturing Methods and Products.

Compression molding	wiring devices, closures, sheets
Expansion bead molding	ice chests, packaging
Extrusion blow molding	hollow objects, bottles
Extrusion	sheets, rods, tubes and profiles
Fluidized bed	plastics coated metal parts
Forging	thermoplastic uniform thick sections
Hand layup	boats, auto bodies, structural sections
Injection molding	thermoset and thermoplastic products
Injection blow molding	bottles and simple shapes
Liquid resin casting	tanks, novelties, encapsulations
Reaction impingement molding	auto bodies and high volume large parts
Rotational molding	tanks, balls, housings, dolls
Spray-up molding	furniture, boats, automobile components
Structural foam molding	business machines, beams, sheets, furniture
Slush molding	novelties, balls, dolls
Sheet thermoforming	
Vacuum forming	blister packages, domes, trays
Pressure forming	furniture, signs, domes
Trapped sheet forming	boxes, machine covers, furniture
Steam pressure forming	ping pong balls, novelties, dolls
Transfer molding	complex thermoset pieces, delicate inserts

surfaces can result when rigid materials are molded. To minimize this problem, the parallel surfaces are tapered slightly in the direction of removal to permit easy breakaway of the surfaces. This taper is referred to as *draft* and is used in almost every plastic molded part. In

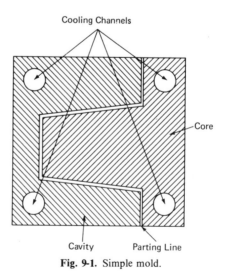

Fig. 9-1. Simple mold.

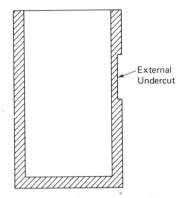

Fig. 9-2. Plastics part with undercut.

the event that two surfaces must be parallel and cannot depart from parallel even by $\frac{1}{4}°$, the mold must be designed so that the sides move apart to permit removal of the piece. Figure 9-3 gives a table of values used to determine draft angles that should be used in designing molded parts. In general, the more generous the draft angle the easier the part will be to mold.

The parting line of a mold, the gate or point of entry of plastics into a mold cavity, and the gas escape vents are three important elements in the design of a molded part. The parting line should be located in a region of the product where the part is most readily molded, usually the perimeter where it is the longest. The parting line, preferably, ought to be straight so that the part can be set down on a plane surface and make contact all around. With slight complications in the mold design and construction, the parting line can be stepped and, in fact, be a very complex mating surface. This can be used to advantage in designing complicated parts which must interlock or nest. Figures 9-4 and 9-5 are examples of lines that are used in molded plastic parts. For more complex designs the mold can be opened in several directions to permit removal of intricate parts. Obviously, the complicated molds will be more costly to build and operate so that the additional product cost introduced should be evaluated as part of the design. In some cases it may be more practical to make two pieces that are subsequently joined than to try to make the plastic part in one piece.

The gate or point of entry of the plastics into a closed mold such as is used in injection and transfer molding is an important point for

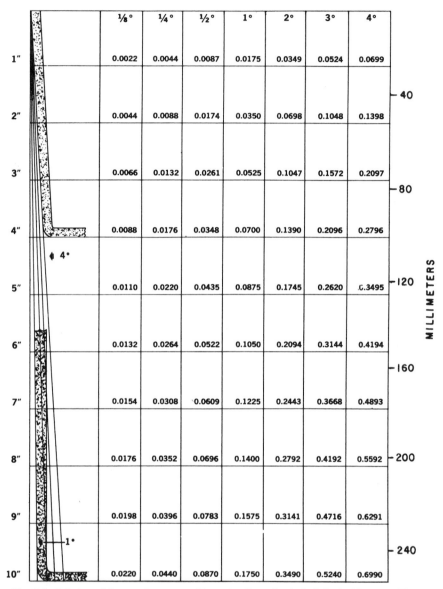

	⅛°	¼°	½°	1°	2°	3°	4°
1"	0.0022	0.0044	0.0087	0.0175	0.0349	0.0524	0.0699
2"	0.0044	0.0088	0.0174	0.0350	0.0698	0.1048	0.1398
3"	0.0066	0.0132	0.0261	0.0525	0.1047	0.1572	0.2097
4"	0.0088	0.0176	0.0348	0.0700	0.1390	0.2096	0.2796
5"	0.0110	0.0220	0.0435	0.0875	0.1745	0.2620	0.3495
6"	0.0132	0.0264	0.0522	0.1050	0.2094	0.3144	0.4194
7"	0.0154	0.0308	0.0609	0.1225	0.2443	0.3668	0.4893
8"	0.0176	0.0352	0.0696	0.1400	0.2792	0.4192	0.5592
9"	0.0198	0.0396	0.0783	0.1575	0.3141	0.4716	0.6291
10"	0.0220	0.0440	0.0870	0.1750	0.3490	0.5240	0.6990

MILLIMETERS

40 — 80 — 120 — 160 — 200 — 240

Fig. 9-3. Relation of degree of taper per side to the dimension in inch/inch. (*Courtesy Plastics Engineering Handbook*)

Fig. 9-4. This molded phenolic valve handle makes use of a stepped parting line, thereby making it unnecessary to provide extremely close alignment between cavity and plunger and facilitating inexpensive cleaning. Note that the attachment insert is held in a molded recess.

the designer to consider, and to specify on the part. The gate must be machined or otherwise fabricated to remove excess material to prevent a noticeable blemish on the part. In general, the gates are made as small and inconspicuous as possible, but this is limited by the properties of the plastics in-flow as well as the filling requirements of the mold. Careful selection of gate location to regions of minimum interference with part function and appearance, compatible with the molding requirements, is necessary. Figure 9-6 shows some typical gate types and locations on molded parts. In many cases injection molds for thermoplastics may be designed for valve gating or hot edge gating which results in no gate stub at all. Hot runner or runnerless molds have done much to eliminate gating problems.

Gate location should be considered from another standpoint. The point of entry of plastics into a mold is a region where plastics flow at very high relative rates and the material in the region of the gate is frequently in a state of frozen in-stress and is usually highly oriented. As a consequence, the material in this region is weakened and is more prone to structural failure by direct stress application, impact, or

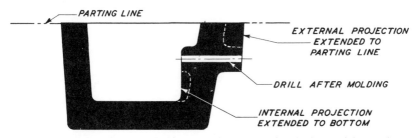

Fig. 9-5. Redesign of piece to facilitate molding. Dotted lines indicate original design.

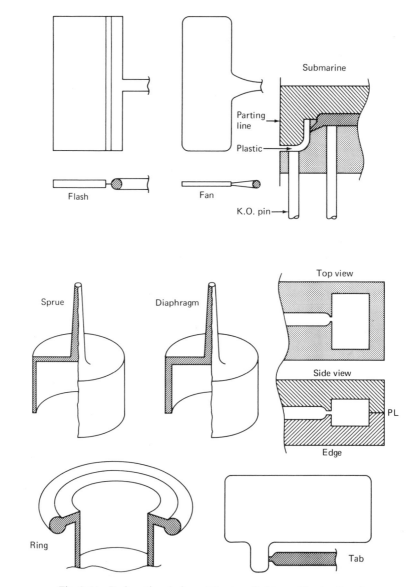

Fig. 9-6A. Basic gating designs. (*Courtesy Robinson Plastics Corp.*)

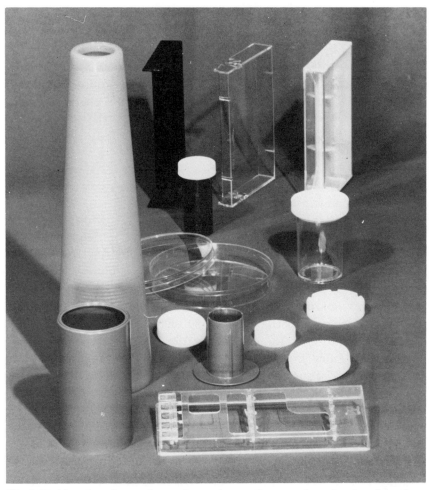

Fig. 9-6B. These hot edge gated parts show only a small gate mark on the edge of the piece.

environmental attack. This should be kept in mind in the part design. If the gate must be located in a high stress area, the part should be suitably strengthened in that region. If possible, the gate should be in a low stress area of the part. Mold designs that facilitate low pressure fill minimize this problem.

The product designer's principal concern with location of the gate is that it may effect the appearance or environmental properties of the product. It is advisable to confer with the mold designer before the product design is finalized. A poorly located gate may result in lines of weakness resulting from failure to knit lines where the pieces will be weak electrically, mechanically, and chemically. Injection mold designs using hot runner molds with hot edge or valve gating will minimize flow lines and gate marks. Thick sections may require very heavy gates. Annular gates are often used on annular sections to eliminate failure to knit lines in the part. Knit lines weaken the properties of the product and failure will result in any pronounced knit line that is stressed.

For a complete study of gating considerations and gate designs, consult DuBois and Pribble.*

Gating for reaction injection molding requires laminar flow and a built in after mixer as part of the gate design. Structural foam and RIM molds require nonconventional gating and venting considerations and the mold designer and the molder should be consulted.

Dimensional tolerances on molded parts are an important area for the designer to consider. Like most manufacturing processes, parts produced by molding will range from the nominal size by an amount which is determined by the process and the materials used. The main source of variation in the case of plastics is the material shrinkage variation. Plastics molding materials shrink as they undergo the change from a viscous fluid to a solid and the shrinkage can range from less than 0.001 inch/inch to as much as 0.050 inch/inch. The shrinkage is affected by the molding pressure and other process variables, and generally varies in different directions in the mold. The variation is usually a percentage of the total shrinkage and, if accurate sizes are critical to a particular application, one of the criteria in material selection must be the mold shrinkage of the material. Complex shapes sometimes necessitate prototype shrinkage measurements to facilitate final mold design.

Figures 9-7 and 9-8 show the type of data** that are available to the product designer for use to set tolerances on injection or compression molded parts. The standard part shown has dimensions in all of the typical directions encountered in molded parts. The graphs show the expected variation in dimension as a function of the part size and within the tolerance level of coarse, standard, and fine. These are

---

Plastics Mold Engineering, Third Edition, Van Nostrand Reinhold Co., New York.
**Standard Practices of Plastics Custom Molders, Published by The Society of the Plastics Industry Inc.

STANDARDS AND PRACTICES OF PLASTICS CUSTOM MOLDERS

Engineering and Technical Standards

GENERAL PURPOSE PHENOLIC

NOTE: The Commercial values shown below represent common production tolerances at the most economical level. The Fine values represent closer tolerances that can be held but at a greater cost.

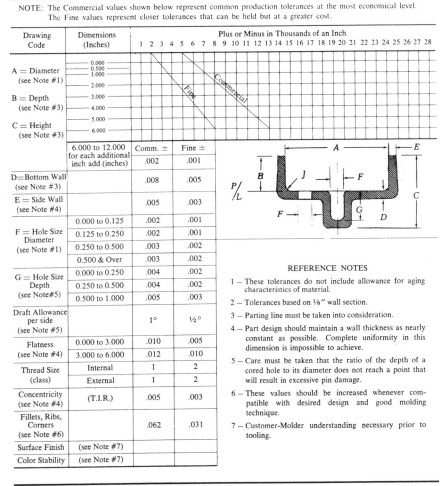

Drawing Code	Dimensions (Inches)		
A = Diameter (see Note #1)	0.000 0.500 1.000 2.000		
B = Depth (see Note #3)	3.000 4.000		
C = Height (see Note #3)	5.000 6.000		
	6.000 to 12.000 for each additional inch add (inches)	Comm. ± .002	Fine ± .001
D=Bottom Wall (see Note #3)		.008	.005
E = Side Wall (see Note #4)		.005	.003
F = Hole Size Diameter (see Note #1)	0.000 to 0.125	.002	.001
	0.125 to 0.250	.002	.001
	0.250 to 0.500	.003	.002
	0.500 & Over	.003	.002
G = Hole Size Depth (see Note#5)	0.000 to 0.250	.004	.002
	0.250 to 0.500	.004	.002
	0.500 to 1.000	.005	.003
Draft Allowance per side (see Note #5)		1°	½°
Flatness (see Note #4)	0.000 to 3.000	.010	.005
	3.000 to 6.000	.012	.010
Thread Size (class)	Internal	1	2
	External	1	2
Concentricity (see Note #4)	(T.I.R.)	.005	.003
Fillets, Ribs, Corners (see Note #6)		.062	.031
Surface Finish	(see Note #7)		
Color Stability	(see Note #7)		

REFERENCE NOTES

1 — These tolerances do not include allowance for aging characteristics of material.

2 — Tolerances based on ⅛" wall section.

3 — Parting line must be taken into consideration.

4 — Part design should maintain a wall thickness as nearly constant as possible. Complete uniformity in this dimension is impossible to achieve.

5 — Care must be taken that the ratio of the depth of a cored hole to its diameter does not reach a point that will result in excessive pin damage.

6 — These values should be increased whenever compatible with desired design and good molding technique.

7 — Customer-Molder understanding necessary prior to tooling.

Fig. 9-7. Standard tolerance chart. (*Courtesy SPI*)

industry accepted levels and the shift from one level to a higher level usually represents an increase in cost of the part and mold as a result of closer control over the molding process, longer molding cycles, and higher reject rates. Alternate materials may be needed.

STANDARDS AND PRACTICES OF PLASTICS CUSTOM MOLDERS

Engineering and Technical Standards

POLYPROPYLENE

NOTE: The Commercial values shown below represent common production tolerances at the most economical level. The Fine values represent closer tolerances that can be held but at a greater cost.

Drawing Code	Dimensions (Inches)	Plus or Minus in Thousands of an Inch
A = Diameter (see Note #1)		
B = Depth (see Note #3)		
C = Height (see Note #3)		

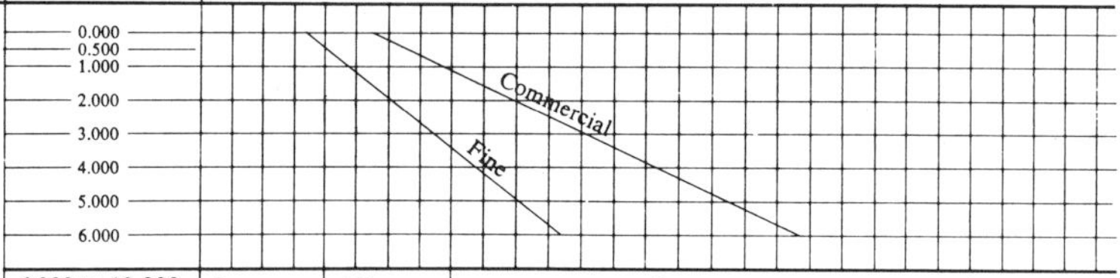

Drawing Code	Dimensions (Inches)	Comm. ±	Fine ±
	6.000 to 12.000 for each additional inch add (inches)	.005	.003
D=Bottom Wall (see Note #3)		.006	.003
E = Side Wall (see Note #4)		.006	.003
F = Hole Size Diameter (see Note #1)	0.000 to 0.125	.003	.002
	0.125 to 0.250	.004	.003
	0.250 to 0.500	.005	.004
	0.500 & Over	.008	.006
G = Hole Size Depth (see Note#5)	0.000 to 0.250	.005	.003
	0.250 to 0.500	.006	.004
	0.500 to 1.000	.009	.006
Draft Allowance per side (see Note #5)		1½°	½°
Flatness (see Note #4)	0.000 to 3.000	.021	.014
	3.000 to 6.000	.035	.021
Thread Size (class)	Internal	1	2
	External	1	2
Concentricity (see Note #4)	(T.I.R.)	.016	.013
Fillets, Ribs, Corners (see Note #6)		.028	.015
Surface Finish	(see Note #7)		
Color Stability	(see Note #7)		

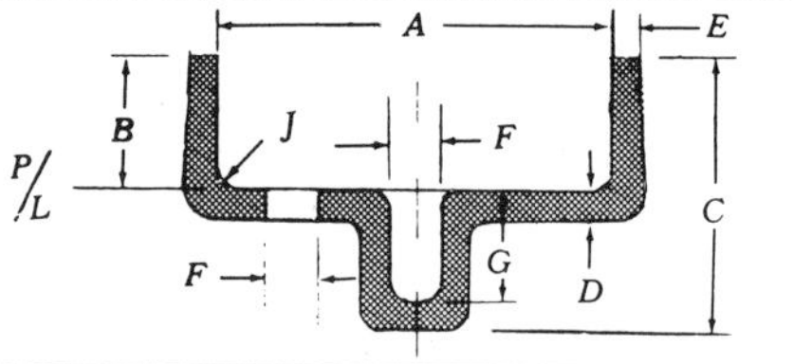

REFERENCE NOTES

1 – These tolerances do not include allowance for aging characteristics of material.

2 – Tolerances based on ⅛" wall section.

3 – Parting line must be taken into consideration.

4 – Part design should maintain a wall thickness as nearly constant as possible. Complete uniformity in this dimension is impossible to achieve.

5 – Care must be taken that the ratio of the depth of a cored hole to its diameter does not reach a point that will result in excessive pin damage.

6 – These values should be increased whenever compatible with desired design and good molding technique.

7 – Customer-Molder understanding necessary prior to tooling.

Fig. 9-8. Standard tolerance chart. (*Courtesy SPI*)

The charts do not reflect the best that can be done with the molding processes. Some precision molders using advanced pressure control methods and very precise control over the plastics melt conditions have consistently produced parts to very good accuracy: better than

0.001 inch variation in 2 inches. Obviously, the cost of producing these parts is much higher than that of conventionally molded parts. In general, as with most manufacturing processes, closer tolerances will greatly increase the manufacturing costs. Where close tolerance parts are required, material selection is important and low shrinkage materials should be used. Some materials such as urea formaldehyde which continue to shrink for several years should be avoided. For very high accuracy parts the product designer should work closely with a molder experienced in precision molding to determine the best design, incorporating the specific experience of the molder as to the dimensions that can be held.

Molded plastics parts in many ways seem analogous to castings since the mold can be any shape. There are a number of limitations on the part because of the fact that the mold is filled with plastics material and it represents, in addition to the form for the product, the flow path for the plastics material. One factor that is affected is the wall thickness of the part. The viscous melts produced by many plastics require that a minimum wall thickness be used based on the ability to fill the mold rather than on the function of the part. For example, rarely is any injection molded part of any size designed with a wall thickness less than 0.025 to 0.030 inch. This is particularly true if the part has a fairly large surface area because the plastics will not flow to properly fill the part. In addition, as will be seen in Chapter 10, flow conditions can severely affect the properties of the material.

Another limitation on molded parts that should be adhered to as closely as possible is to design with uniform wall thickness. The molding processes involve heat exchange and thicker walls cool at substantially lower rates than thin walls. One of the consequences of having parts with widely varying wall thickness is potential postmolding distortion of the parts. In addition, the molding cycle time is limited by the cooling rate of the thickest section so that having even one small thick area will result in a longer and more expensive molding cycle.

Another limitation on wall thickness consideration for thermoplastic is the effect that it has on molding blemishes called *sink marks* where the surface of the plastics part develops a dished-in area that looks like a finger depression. Thick sections, particularly thick sections of varying thickness, are most likely to cause sink marks on the part. It is true that the molding conditions determine the extent of the sink mark defect; it is equally true that some part designs are impossible to mold without sink marks because of the design.

Ribs and bosses are frequently designed into plastics parts to supply stiffness and to support specific areas. Unless these are designed with the shrinkage characteristics of the plastics in mind, they can result in surface blemishes on the opposite side of the walls to which they are attached. Figures 9-9 and 9-10 show some typical rib and boss configurations with suggested dimensions to minimize the shrinkage defects. External flutes or reeds are appearance design treatments that compensate for thick sections and hide sinks.

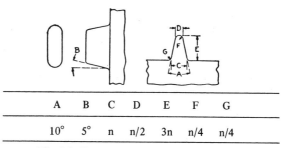

A	B	C	D	E	F	G
10°	5°	n	n/2	3n	n/4	n/4

Fig. 9-9. Recommended proportions for ribs.

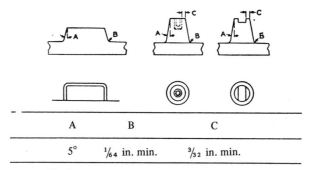

A	B	C
5°	¹⁄₆₄ in. min.	³⁄₃₂ in. min.

Fig. 9-10. Recommended proportions for bosses.

Plastics tend to exhibit notch sensitivity so that sharp corners which are stress risers should be avoided. Figure 9-11 which is widely reprinted in all material on plastics design shows graphically the effect of sharp corners on the stress concentration in molded parts. The photograph shows a photo-stress figure using the effect on polarized light of stressed materials as discussed in previous chapters. It

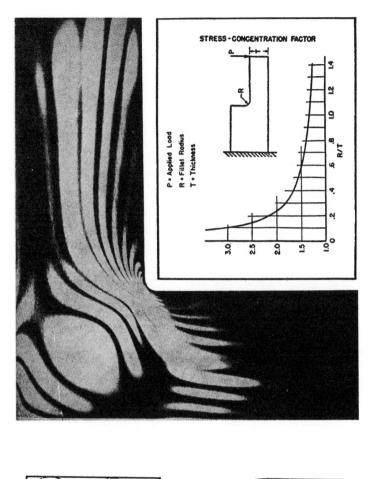

Fig. 9-11. (*Courtesy Plastics Engineering Handbook, New York, Van Nostrand Reinhold, 1960*)

is apparent from Fig. 9-11 that exterior sharp corners also contribute to increased stress, and it is recommended that wherever practical both inside and outside corners should have generous radii.

Molded parts can have a variety of surface treatments that range from high gloss surfaces to sandblasted effects to fluting and patterning of many kinds simulating wood, leather, cloth, and other natural materials. These surface effects may be used to enhance the appearance of plastics parts and to mask out surface blemishes such as sink and flow marks. The patterning should be designed on the surface so that it will not complicate ejection of the part from the mold. Deep surface graining on a nearly vertical surface may represent enough of an undercut to make parts stick in the mold. Despite this limitation, it is usually easy to use one or more of the available surface treatments to improve the appearance of a molded part.

Holes through the walls and into the sections of a molded part re-represent a design problem that must also be considered. The holes through the wall of a molded part are made by core projections in a mold. If the holes lie in a plane parallel to, or nearly parallel to the parting line of a mold, they are easy to make and do not complicate the design of the tool. If the holes are on a surface that is perpendicular to the parting line of the mold, they require that the cores be moved out of the way before the part can be removed from the mold. Holes can be punched in many thermoplastic parts. Figure 9-12 shows one way in which side holes can be designed into a part and still not require the use of retractable cores. The sliding cut-off is not

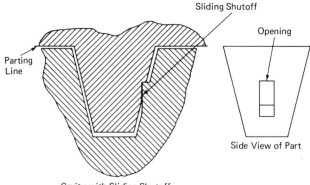

Fig. 9-12. Cavity with sliding shutoff.

the easiest type of mold to generate but it is substantially less costly and complicated than retractable cores. Jiggler pins may be used for internal side wall holes as shown in Fig. 9-13.

The location of holes through the walls of a molded part is important relative to the gate location and the flow path of the material. The holes represent dams across the flow path of the material that it must flow around and then rejoin on the opposite side. This restriction in flow can lead to poor flow and weld lines in the material where the material rejoins. These weld areas tend to be weak spots in the

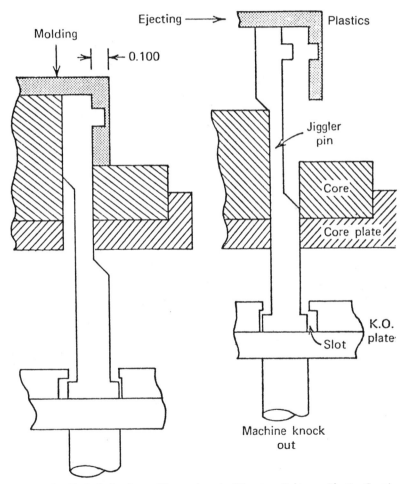

Fig. 9-13. "Jiggler" pins for molding undercots. (*Courtesy Robinson Plastics Corp.*)

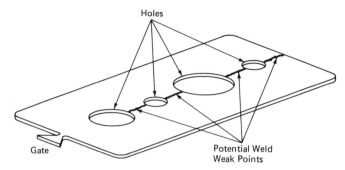

Fig. 9-14. Problem molded part.

molded part and the flow pattern should be adjusted so that they occur in noncritical portions of the part. Figure 9-14 shows a molded part with a series of holes that can cause flow and weld problems. Relocation of the gate can improve the flow and move the weld to a less critical place on the part. Holes are often molded two thirds of the way through the wall only with final drilling to eliminate weld lines. Inserts as shown in Figure 9-15 are molded into plastic parts for design integration. Standard design inserts are available for inclusion. The insert section must be appropriate to the plastics section to avoid differential expansion breakage.

Holes used for inserting screws, pins, or other fasteners to join molded parts to other pieces are frequently used. They can be molded in or machined as secondary operations. The choice depends on the accuracy required in location and size, as well as the potential complications that molding in the holes will produce in the mold and the molding process. For example, deep holes which are perpendicular to the side walls and require a long side draw core should, in many cases, be drilled. Holes perpendicular to the parting line, particularly holes terminating on the parting line, should be molded in except when they are very deep or must be located with a high degree of accuracy. Long hole products should be designed so that the mold pin can enter the opposite half of the mold to gain support at both ends. Figure 9-16 depicts methods used to eliminate many molded product problems.

These are some of the considerations that the designer must make in the design of a molded part. *Plastics Mold Engineering** gives a

---

*DuBois and Pribble, *Plastics Mold Engineering*, 3rd Edition, New York, Van Nostrand Reinhold Co. 1977.

BLIND HOLE   **OPEN HOLE**   **BLIND HOLE
COUNTERBORED**

**BLIND HOLE
PROTRUDING**   **EYELET
PROTRUDING**   **EYELET
BOTH ENDS
PROTRUDING**

**PROTRUDING
RIVET**   **DOUBLE
PROTRUDING
WITH THREADS**   **PROTRUDING
EYELET
WITH INTERNAL
THREADS**

DRAWN PIN   **DRAWN SHELL**   **DRAWN EYELET**

Fig. 9-15. Usual types of inserts. (*Courtesy SPI*)

more detailed discussion of the mold limitation factors for many specific problems encountered in product design practice. In all cases the molder should be made a party to the design of a molded part so that complications that may result from the molding can be anticipated and designed around.

Each of the plastics processes has die and processing limitations that restrict freedom of design. Molding has the most well defined set of restrictions. It is possible to mold any given shape if the product

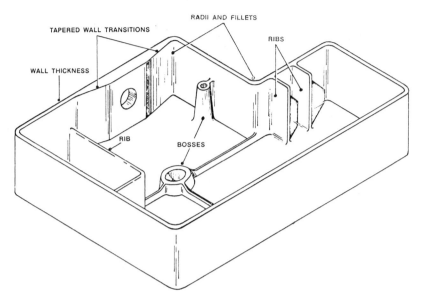

Fig. 9-16. This sketch illustrates many features of a well designed part and points out problem areas. (*Courtesy Marbon Div. Borg Warner Corp., Washington, W. Va.*)

can pay the price for mold and molding. Extrusion, blow molding, vacuum forming, and casting have wide use. The design limitations imposed by these processes will be briefly discussed.

Extrusion is a process where a plastics material is softened and flowed through a shaped orifice after which it is cooled to produce a continuous shape with a constant profile. Examples of extruded shapes are pipe and tubing, sheets, strips, rods, and profiles such as an angle or channel section. The process is frequently used with an in-line secondary operation where the part may be cut to length, have holes punched in, or surface grained or some similar treatment so that a more nearly complete part is produced. Figure 9-17 shows some examples of products made by extrusion. Extrusion is a very versatile method of processing. It can be used to produce composite materials, coat other material, and even produce cellular material shapes. Our discussion will include shapes since the design of specific composites, covering, etc., would not normally be a product design engineering activity.

Basic stock shapes such as rod, tube, strip, and sheet in most plastics materials are available as off-the-shelf items much as similar

Fig. 9-17A. Simulated wood products extruded from a structural foam with bonded wood print. (*Courtesy Anchor Plastics Co., Long Island City, N.Y.*)

shapes in metals. They generally are used as raw materials in subsequent fabrication operations like machining and forming. Special shapes such as an edge channel to cover the edge of a metal piece in an appliance is a custom shape and thus extruded to the designer's requirements. Figure 9-18 shows a typical shape that might be used to form a special joining unit in a business machine. It has several of the typical sections that are required in such extrusions and will be used to define the design problems that result from process limitations. One of the important characteristics of extruded parts is that the section represents the effect of flow through an orifice which generally is a

Fig. 9-17B. Examples of sheathed structural moldings with particle board and plywood core sections. (*Courtesy Shock D-706, Schomdorf, W. Germany*)

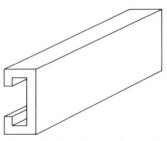

Fig. 9-18. Extruded plastics profile.

plate with the desired cross section or a modification of it to compensate for shape changes caused by the material flow characteristics. The easiest flow to predict is one in which all of the walls of the part are of equal thickness, so one desirable design feature in an extruded part is that it have uniform walls. If this is not possible, the next best thing is to have discrete steps in wall thickness. The least desirable shapes in terms of flow control are varying wall thicknesses. The more complicated wall thickness configurations require extensive adjustment of the tool to produce the part. Generally, such a part can only be produced from a limited number of materials with melt characteristics that exhibit a high degree of melt stiffness and thixotropy. Two of the best materials to extrude complex shapes from are rigid polyvinyl chloride and ABS materials. One thing to keep in mind is that a die that has been corrected to run one material to the required shape may not run another material, even another formulation of the same basic resin, and will generally not give the correct shape in another machine with different screw and barrel characteristics. Unlike molding tooling, it is very difficult to transfer extrusion tooling from one plant to another. In addition to the machine characteristic changes, each plant has different post-extrusion handling jigs and fixtures which complete the shaping of the profile. This will also have a substantial effect on the shape of the resulting part.

Dimensions can be held most closely between sections attached to the same element. For example, the dimensions of a channel section will be accurate at the bottom of the channel where they are set by the spacing in the die. The upper portion of the channel at the end of the legs is more difficult to control since the cooling of the section usually has a tendency to make the walls bend inward. To minimize this tendency the section is usually made over-compensated with the walls bent slightly out so that the distortion will bring them back into proper position. The jigs used in extrusion are also used to over-compensate the shape so that the cooled shape falls within the required dimension.

Certain shapes are very difficult to extrude because of the intrinsic flow properties of plastics. A square cross section with sharp corners is one of the most difficult shapes. The shrinkage, plus flow patterns, can cause the shape to vary from a barrel version to a concave side version of a square as shown in Fig. 9-19. The material used has a

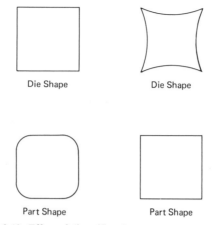

Fig. 9-19. Effect of die orifice shape on square extrusion.

major influence on this. It is best to avoid this type of shape and it is good design practice with extrusions to use generous internal and external radii on all corners. If a square shape is required, it may be best to use a square tube that can be held to shape by the use of appropriate shaping jigs.

One of the characteristics of the extrusion process is that when the material extrudes from the die there is some swelling of the section. After the material exits the die it is usually stretched or drawn down to a size equal to or smaller than the die opening. The dimensions are reduced proportionally so that in an ideal material the drawn down section is the same as the original section but smaller proportionally in each dimension. Because of melt-elasticity effects on the material it does not draw down in a simple proportional manner so that the drawn-down process is the source of errors in the profile. These must be corrected by modifying the die and the forming jigs.

There are substantial influences on the material caused by the extrusion so that the properties are different in the flow direction and perpendicular to the flow direction. These have a significant effect on the performance characteristics of the part and will be discussed in Chapter 10.

Blow molding is another important process for making plastics articles, particularly hollow container shapes such as bottles. In this process a tubular element of hot plastics is captivated in a mold which has the external shape of the part desired. Air is introduced into the

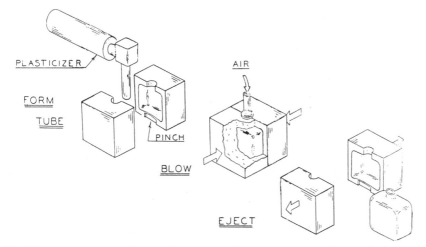

Fig. 9-20. Basic pinch-tube blow-molding process. (*Courtesy Monsanto Co., Hartford, Conn.*)

interior of the hot plastics element called the *parison* and the plastics is blow-expanded to conform to the wall of the mold. Figure 9-20 shows the process schematically. In general, the only area of the blown part that has accurately controlled walls and external shapes as well, is the neck where the parison is gripped and where the air is usually introduced. The wall thickness of the rest of the part will depend on the shape of the mold, the wall thickness of the parison in the different sections, the characteristics of the plastics, and the blowing temperature. By programmed parison design it is now possible to make any bottle shapes with fairly uniform walls even when the bottle is square rather than round and when the periphery of the shape is large compared with the length. Blow molded products (Fig. 9-21) are produced by an extrusion blow process (Fig. 9-20) and an injection blow process (Fig. 9-22).

Industrial shapes, tanks, gasoline tanks, vehicles and hospital ware such as urinals are blow molded. These shapes require complex blowing patterns because of the odd and nonsymmetrical shapes involved. The parison thickness control is difficult and usually the wall thickness on these parts is not as uniform as on the symmetrical bottle shapes. In designing parts for this process which have specific wall thickness requirements for stiffness and strength, it is advisable to use heavier walls because of the possibility that control in the blow mold-

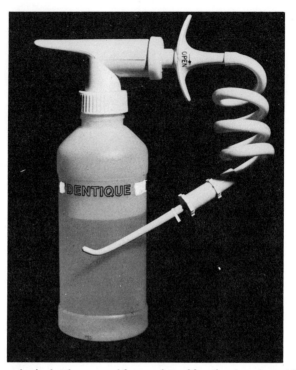

Fig. 9-21. Blown plastics bottles are used for a variety of functional products. (*Courtesy Dentique, Bethesda, Md.*)

ing process is insufficient to avoid thin spots in locations where the material will be subjected to high stress levels in service.

The only general rule in designing parts by blow molding is to avoid sharp changes in direction on a part. Reentrant shapes are particularly difficult to blow mold. The surface configurations should be simple. Blend all surfaces together and use generous radii at any bends. The wall thicknesses must have a wide tolerance because it is difficult to control except in the simplest shapes. There must be a neck section on each part where the blow tube enters the part. In this section the dimensions are controlled between the blow stick and the mold so that this part is essentially injection molded and the dimensions are easier to control. In extrusion blow molded parts there is a place where the parison tail is removed. This is an area where there is a fold and it is generally a weak section on the part. It is usually at the bottom of a bottle shape. When its location is on a more complex shape it should

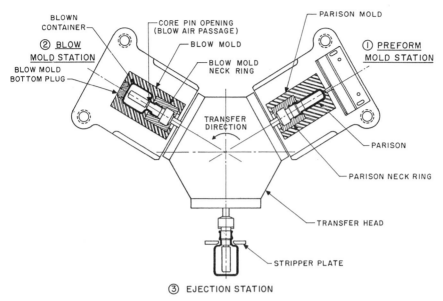

Fig. 9-22A. Schematic of three-station rotary press.

Fig. 9-22B. Injection blow molding machine.

be carefully considered to minimize the interference with the part function; a "push up" producing a depressed area is commonly used as a cure.

As in the case of the other processing techniques for the plastics, the processing has a pronounced effect on the environmental stability and physical characteristics of the part and the effect of these on product performance is discussed in the next chapter.

Reaction injection molding (RIM), also called high pressure impingement molding, is used for automobile bumbers, fascia, fenders, etc. where a rigid but elastic part is needed (Fig. 9-23). In this process, the two components of a resin such as urethane are metered carefully and mixed at very high pressure in a mixing chamber prior to injection into the mold where a fast thermoset cure is achieved. A similar metering mixing system is used for cast urethane products.

Many of the polymers such as caprolactam, silicone, urethane, and polyester are multicomponent reactive liquids that may be molded by the reaction injection process which is particularly suitable for large area pieces such as auto body parts, furniture, and housing sections, because of the fast cycle and low mold clamp pressure requirement. This system greatly diminishes the capital investment requirement for such large parts. A 5000-ton conventional injection press is presently the largest and it costs $1,000,000. An automobile fender would need a 12,000 ton press costing $2,000,000 by the conventional injection process; with reaction injection equipment cost might be one tenth of the machine cost for the older process.

Vacuum forming or thermoforming processes used to shape plastics sheet materials is one of the larger areas of production for plastics parts. The beginning material is a thermoplastic sheet, usually made by extrusion, that is heated to an appropriate forming temperature. In some cases the sheet is formed directly as it leaves the extruder, but in most cases it is reheated in the forming machine. Figure 9-24 shows some of the forming operations used but in each case the heated sheet is taken over a male or female mold and stretched so that it conforms to the shape of the mold. This is done by the application of pressure or vacuum, or both. The wall thickness of the resulting part is usually less than that of the starting material in all areas but, in any event, it is less in most locations because of the stretching which takes place during the forming operation.

By the nature of the operation, the formed part must be such that

Fig. 9-23. Reaction injection molding is used for non-rigid large components.

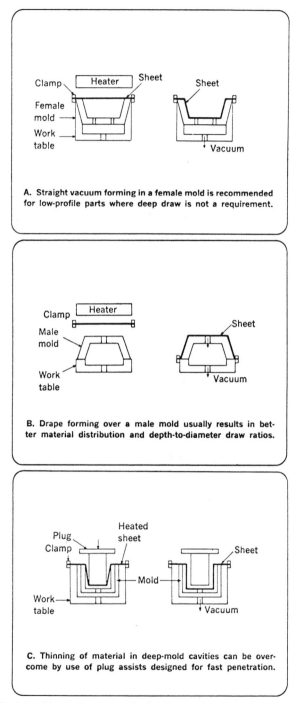

A. Straight vacuum forming in a female mold is recommended for low-profile parts where deep draw is not a requirement.

B. Drape forming over a male mold usually results in better material distribution and depth-to-diameter draw ratios.

C. Thinning of material in deep-mold cavities can be overcome by use of plug assists designed for fast penetration.

Fig. 9-24. Thermoforming methods. (*Courtesy Mobay Chemical Co.*)

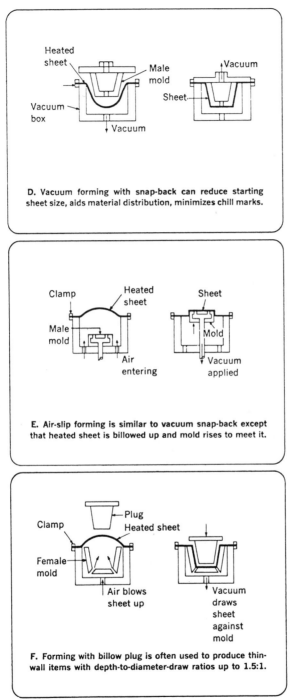

D. Vacuum forming with snap-back can reduce starting sheet size, aids material distribution, minimizes chill marks.

E. Air-slip forming is similar to vacuum snap-back except that heated sheet is billowed up and mold rises to meet it.

F. Forming with billow plug is often used to produce thin-wall items with depth-to-diameter-draw ratios up to 1.5:1.

Fig. 9-24. (*Continued*)

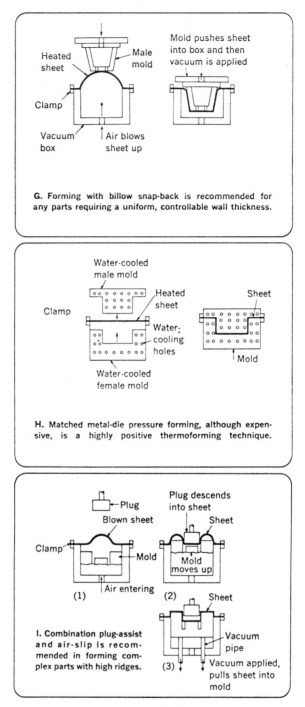

G. Forming with billow snap-back is recommended for any parts requiring a uniform, controllable wall thickness.

H. Matched metal-die pressure forming, although expensive, is a highly positive thermoforming technique.

I. Combination plug-assist and air-slip is recommended in forming complex parts with high ridges.

Fig. 9-24. (*Continued*)

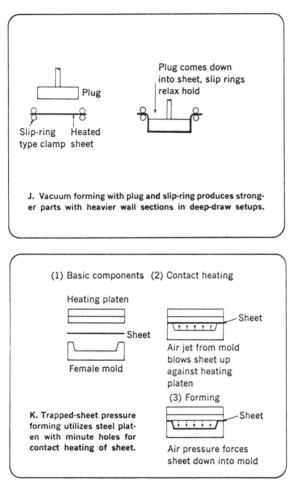

J. Vacuum forming with plug and slip-ring produces stronger parts with heavier wall sections in deep-draw setups.

K. Trapped-sheet pressure forming utilizes steel platen with minute holes for contact heating of sheet.

Fig. 9-24. (*Continued*)

it can be conformed to a male or female mold without any re-entrant sections that will form undercuts to prevent it from being easy to eject from the mold. In general, thermoformed product molds do not have moveable cores that permit the molding of undercuts. Any holes in the part must be made subsequent to the forming process, usually by a die cutting or punching process. In thicker materials the holes may be drilled or machined with routing cutters. Some thermosetting laminates may be thermoformed by preheating in a molten lead bath or by dielectric heating.

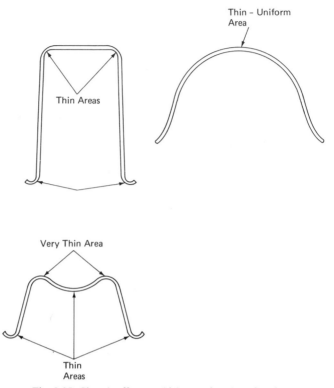

Fig. 9-25. Shape's effect on thickness after sheet forming.

One specific design characteristic of thermoformed plastics parts is that the product has a reasonably uniform wall thickness. It is a sheet forming process and has the same type of design character- on wall thicknesses and shape as a drawn steel part. Sharp corners are difficult in formed plastic parts. The material is usually very thin in sharp corner areas and weak because of the high degree of draw-down to form around the corner (Fig. 9-25). Smooth compound curves make excellent shapes for formed parts and they are usually joined by large flat areas. Reinforcements are corrugations or single corrugation type ribs. Frame-like raised sections make good stiffen- ers, particularly around areas that must be subsequently trimmed. Several of these treatments are shown in Fig. 9-26).

Not all plastics are readily formed and there is a limited range of thicknesses that can be formed. A list of typical vacuum formable materials is given in Table 9-2 along with the range of thicknesses

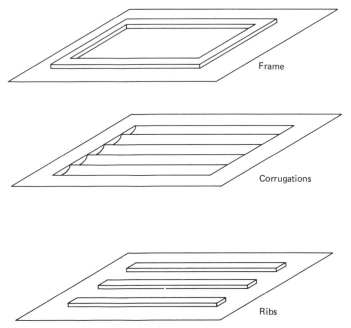

Fig. 9-26. Stiffening designs for sheet formed parts.

that can be formed. Generally, the easily formed materials such as rigid PVC, oriented styrene, and cellulose acetate propionate can be formed from very thin material, down to .005 inch thick. This makes these materials desirable to use for disposable packaging such as trays and bubble packs. ABS material can be formed in thicknesses up to 3/16 inch or more. This material is used to make large parts such as

Table 9-2. Types and Gage of Materials used in Thermoforming

Material Name	Range of Gage (in)
Acrylonitrile butadiene styrene (ABS)	0.005–0.250
Polyvinyl chloride (PVC) rigid	0.002–0.060
Methyl methacrylate	0.030–0.500
Cellulose acetate	0.010–0.060
Cellulose acetate butyrate	0.005–0.060
High impact polystyrene	0.010–0.250
Polystyrene (general purpose)	0.005–0.030
Polyethylene	0.010–0.125

machine housings, automobile bodies, and small boats. Frequently the larger items are reinforced with foam to impart stiffness and, in some cases, the sheet material itself contains a self-blowing core that makes the part a sandwich panel section after forming.

Surface treatment on formed parts is limited, especially as to surface treatments that are done in the forming process. The low pressure used, and the fact that the material is formed at a relatively low temperature compared to the melt temperature, make detail hard to achieve. By the same token, it is possible to start with materials that have a texture embossed on them that will retain the surface texture throughout the forming process.

The sheet forming processes are a good way to make large low-volume and small high-volume products where the sharp detail possible by the molding process is not required. The tooling costs compared with the part size are modest. Frequently the thermoforming process is used in the first stages of manufacture of a product that can use the type of parts produced. The mass production is later converted to the injection or other molding process.

Solid phase forming or forging is a process wherein a heated blank of thermoplastic material is placed in a female die with a male punch and reformed under impact or high compression pressure. Hydraulic or toggle presses are used to form the simple shapes shown in Fig. 9-27. This process is commonly used for thick section parts, for example, wide face gears made of thermoplastics. Such pieces would have a

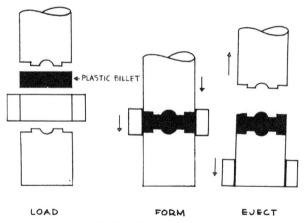

LOAD   FORM   EJECT

Fig. 9-27. Forging process.

Table 9-3. Typical Casting Resin Systems.

Methyl methacrylate casting resins peroxide catalyzed
Polyester casting resin systems peroxide catalyzed metal soap promoted
Epoxy casting resins bisphenol type amine or polyamid catalyzed
Epoxy casting resins aliphatic type organic acid catalyzed
Urethane casting resins tin salt and amine catalyzed
Phenolic resin casting systems acid or amine catalyzed
Vinyl casting systems using plastisol-type dispersion heat converted
Silicone elastomer casting systems using lead peroxide catalyst
Polybutadiene casting resins using organic peroxide catalyst

very long cure in an injection machine. In making these forged parts, blanks are cut from extruded bars or sheets, preheated to a point just below the melt temperature of the plastics prior to forging.

Plastics casting is a process in which a very fluid mixture of a resin composition is poured into a mold and solidified by a chemical or physical chemical change by the application of catalysts, heat, or actinic and ionizing radiation. A number of materials are frequently cast and a list of these is given in Table 9-3. Generally a casting system is used either because the material cannot be processed by other means or the size and nature of the product is such that casting is necessary. It is also used for prototypes in cheap molds.

An example of a plastics systems and product type where casting is chosen is the use of epoxy resins to encapsulate transformers and other electrical and electronic circuitry. The resin permeates the encapsulated part and provides internal electrical insulation so that the entire unit is solidified into a convenient package with the electrical circuitry protected from the environment. Relatively simple molds are used and the resin mixture is poured into the mold containing the part to be encapsulated. Usually catalyzed mixtures are used and heat is produced by the chemical reaction of the catalyzed mixture. When the resin solidifies and cools it shrinks to some degree, depending on the formulation. The solid part will be somewhat smaller than the mold used. The heat used or produced by the cure and the shrinkage represents two design factors that must be taken into account. If the parts being encapsulated are heat-sensitive, such as semiconductor devices, the heat rise must be determined and, by proper formulation, kept to an amount that will limit the casting temperature to one below the critical failure temperature of the encapsulated product. The use of fillers which dilute the reaction mix-

1. Inserts are placed in mold loading rack.

2. Rack is inserted over mold located in press.

3. Press is closed. A measured amount of powdered resin is poured into material pot.

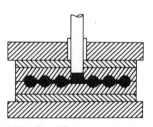

4. Application of molding pressure forces plasticized resin through runners and gates to die cavities, encircling components.

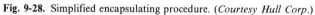

Fig. 9-28. Simplified encapsulating procedure. (*Courtesy Hull Corp.*)

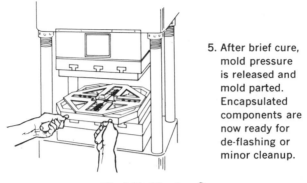

5. After brief cure, mold pressure is released and mold parted. Encapsulated components are now ready for de-flashing or minor cleanup.

Fig. 9-28. (*Continued*)

ture, as well as the use of slow, less active, catalysts are the two routes generally taken to control heat rise. Many electrical and electronic components are encapsulated by transfer molding as shown in Fig. 9-28.

The shrinkage in conjunction with the mechanical properties of the cured resin is an important consideration in encapsulating delicate structures such as integrated circuits and ceramic capacitors. If the resin hardens to a very rigid material and has a fairly high shrinkage, it can crush the embedded part. Resins which harden to a semirigid

Fig. 9-29. Coil with universal inserts encapsulated by transfer molding.

mass are preferred and the shrinkage is controlled by the use of a filler or a catalyst or hardener which causes minimum shrinkage. It is fortunate that the requirements of shrinkage control are essentially the same as those for heat control so that good encapsulating formulations are generally available. The only compromise required is that the filler generally increases the viscosity of the formulation and it is necessary to limit the amounts used when intricate configurations involving small flow paths are encapsulated. Figure 9-29 shows some examples of castings of this type.

The second major area for cast products is in such architectural units as simulated marble sink tops, wall placques, and decorative sculpture. For these products the preferred materials are polyester and acrylic recipes which contain large amounts of mineral fiber such as marble dust, clay, and chalk. These parts are usually made by casting into closed molds using fairly low viscosity mixtures with room temperature curing catalysts. Large parts measuring up to 8 feet square with thicknesses ranging from 1/16 to 1/2 inch thick can be made and

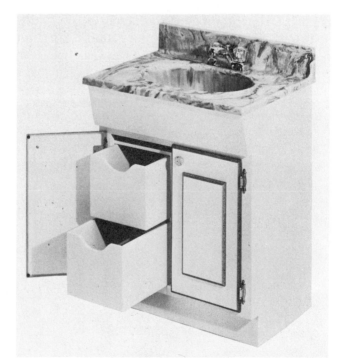

Fig. 9-30. The top of this sink is marble filled polyester. (*Courtesy General Marble Co.*)

the physical properties of the castings are usually quite good. Figure 9-30 shows some examples of cast sink units. The design for products of this type would parallel that of cast metal products into precision molds. Parting line and venting considerations would be made and material selection would be another engineering consideration. For example, a material used for a sink top sould be resistant to soap and cosmetic agents and household chemicals. The material should have good impact and chip strength and be fairly abrasion resistant to hold up under normal cleaning.

The third area for casting is the use of cast elastomer products, particularly for industrial and medical products. Polyurethane and polyester based formulations are generally used, as is the room temperature vulcanizing silicone (RTV) material. These formulations can be cured at room temperature or at low oven temperatures and they

Fig. 9-31A. Ultrasonic bonding of thermoplastic products is adaptable to complex high volume products as well as for simple pieces. Shown here is a semiautomatic slide table ultrasonic assembly system that uses standard welding heads and power supply to weld polystyrene automotive defroster duct assemblies. The high frequency vibration generated by the head generates heat at the weld line for bonding. Fast welds are achieved by this process. (*Courtesy Sonics & Materials, Inc.*)

represent a low tooling cost approach to the manufacture of rubber-like parts. The RTV silicones are widely used to make molds for the casting of other resins. A master can be reproduced in exquisite detail using the appropriate RTV compound for making a mold that can be used subsequently to cast a polyester, an expoy, or a polyurethane part. The RTV is a self-release material to the other resin which simplifies the casting procedure. The polyurethane products usually produced are shock absorber assemblies, rollers and drive wheels, and caster tires. The polyester materials are used for similar purposes. The silicone materials are widely used in medical prosthesis applications and, with special materials, for surgical implants.

Design limitations are primarily involved in material selection to produce the desired properties with appropriate wall thicknesses. The materials range in shrinkage during cure, from practically zero for

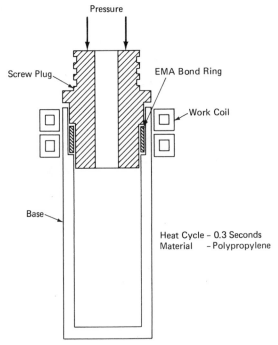

Fig. 9-31B. Electromagnetic bonding of thermoplastics is achieved by use of an electromagnetic energy absorbing material in a thermoplastic of the same composition as the parts being bonded. A high frequency induction heating device is employed to supply energy to the work coils that generate the heat in the bonding composition (EMABond Ring). High pressure and vacuum seals are achieved easily in production procedures. (*Courtesy EMABond, Inc.*)

the RTV silicone materials, to 1 to 3% for the other materials. Other than the allowance for this shrinkage, there are few constraints on shapes. As a matter of fact, because of the rubbery nature of the materials, undercuts are readily tolerated and shapes that would require complex molds in a rigid material can be readily done with simple molds.

Parts are also made from thermoplastic and thermoset materials by machining from rod, bar, sheet, and tube. They are often forgotten or overlooked by designers. A punched laminate is cheap. The machining operations are similar to woodworking and metal working modified with special cutters, drills, and other tools for the machining characteristics of specific plastics. Heat generation is a problem and special cooling is required. Suppliers of the basic stock provide machining recommendations for the material. Good practice is essential to the production of parts that will not fail from machining stresses and errors.

Good adhesives are available for bonding the plastic materials. Ultrasonic and electromagnetic bonding as shown in Figs. 9-31A and 9-31B give excellent results and should be evaluated for each job. Other bonding methods include the use of epoxy adhesives for thermo-

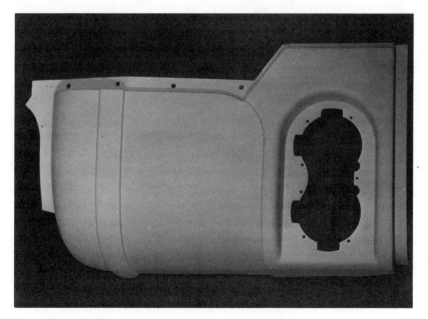

Fig. 9-32. Front quarter panel of a truck body molded of polyester glass.

sets, solvents for thermoplastics, fusion and frictional heat. Hot gas torches and hot plates are often used in the welding of thermoplastics.

The reinforced plastics are molded, extruded, laminated, compression, transfer or injection molded or formed by simple hand or machine layup processes. The reinforced plastics are commonly referred to as those parts made from a polyester or glass epoxy combination as opposed to those that are made from resin-bonded fillers molded in the conventional manner. Polyester glass bulk molding compounds (BMC) are compression molded in low pressure presses for high strength products as shown in Fig. 9-32. Sheet molding compounds (SMC) are also used for parts of relatively uniform thickness by loading preforms of uncured sheets made up of glass fibers and a reactive resin. Other reinforced plastics molding methods are shown in Figs. 9-33 to 9-40. Product designs for these pieces are simple and few limitations are placed on the designer. It is preferable to drill holes after molding when the transfer process is used, to insure maximum strength. Large quantities of these high strength sections are used for automobile components and boats.

The constraints on design imposed by the processes for manufacture have been covered in this chapter for a number of the widely used plastics manufacturing processes. The first step in product design should be the selection of the material and manufacturing process. Not all processes were covered, just those that are most frequently met in design practice. The basic process limitations were discussed

CONTACT MOLDING

Resin is in contact with air. Lay-up normally cures at room temperature. Heat may accelerate cure. A smoother exposed side may be achieved by wiping on cellophane.

Fig. 9-33.

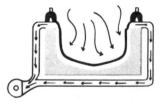

VACUUM BAG

Cellophane or polyvinyl acetate is placed over lay-up. Joints are sealed with plastic; vacuum is drawn. Resultant atmospheric pressure eliminates voids and forces out entrapped air and excess resin.

Fig. 9-34.

Figs. 9-33 through 9-40 *Courtesy Owens Corning Fiberglass Corp.*

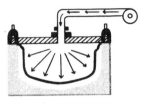

AUTOCLAVE

PRESSURE BAG

Tailored bag – normally rubber sheeting–is placed against lay-up. Air or steam pressure up to 50 psi is applied between pressure plate and bag.

Fig. 9-35.

Modification of the pressure bag method: after lay-up, entire assembly is placed in steam autoclave at 50 to 100 psi. Additional pressure achieves higher glass loadings and improved removal of air.

Fig. 9-36.

SPRAY-UP

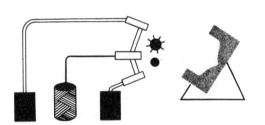

Fiberglas and resin are simultaneously deposited in a mold. Roving is fed through a chopper and ejected into a resin stream, which is directed at the mold by either of two spray systems: (1) A gun carries resin premixed with catalyst, another gun carries resin premixed with accelerator. (2) In a second system, ingredients are fed into a single gun mixing chamber ahead of the spray nozzle. By either method the resin mix precoats the strands and the merged spray is directed into the mold by the operator. The glass-resin mix is rolled by hand to remove air, lay down the fibers, and smooth the surface. Curing is similar to hand lay-up.

Fig. 9-37. Spray-up reinforced plastics.

to guide the design. Close cooperation with the plastics processor who will make the product is desirable to minimize complications in tooling and manufacture and to permit the proper level of precision to be maintained. Each specific part has some problems that the processor can help solve. Materials makers literature will provide excellent guidance in product design details. As a result of the pro-

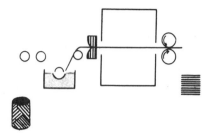

Continuous strand – in the form of roving – or other forms of reinforcement is impregnated in a resin bath and drawn through a die which sets the shape of the stock and controls the resin content. Final cure is effected in an oven through which the stock is drawn by a suitable pulling device.

Fig. 9-38. Continuous pultrusion reinforced plastics.

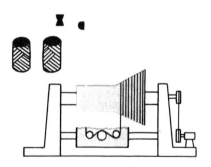

Filament winding uses continuous reinforcement to achieve efficient utilization of glass fiber strength. Roving or single strands are fed from a creel through a bath of resin and wound on suitably designed mandrel. Preimpregnated roving is also used. Special lathes lay down glass in a predetermined pattern to give max. strength in the directions required. When the right number of layers have been applied, the wound mandrel is cured at room temperature or in an oven.

Fig. 9-39. Filament winding process.

PLENUM CHAMBER

Roving is fed into a cutter on top of plenum chamber. Chopped strands are directed onto a spinning fiber distributor to separate chopped strands and distribute strands uniformly in plenum chamber. Falling strands are sucked onto preform screen. Resinous binder is sprayed on. Preform is positioned in a curing oven. New screen is indexed in plenum chamber for repeat cycle.

Fig. 9-40.

cessor's detailed experience or unique equipment he may be able to point out an unknown area that may require experimentation and test prior to production. While pilot tooling and the delay that is attendant on its use is not always justified, the processor can indicate when it is essential. Designing successful plastics parts requires the cooperation of all concerned—the engineer, the stylist, and the processor—so that the product is useful, durable, and economical to produce.

Chapter 10

Material and Process Interaction and the Effects on the Performance of Plastics Parts and the Resulting Design Limitations

When a part is made from a plastics material, the conversion process subjects the material, in most cases, to rather severe physical conditions involving elevated temperatures, high pressures, and high shear rate flow, as well as chemical changes. These drastic processing conditions place some limitations on the design of plastics parts and on their performance. The discussion here will first cover the molding processes of injection, transfer, and compression molding and then other forming processes.

In injection molding the walls and other sections of a plastics part represent to the designer the things required to make the part functional for its intended use. To the mold designer and molder they represent the flow path for the plastics material. This flow takes place at high rates and under the complicating conditions of flow in a passage much cooler than the resin, through an orifice (the gate) whose dimensions are severely restricted to reduce the effect on the appearance of the part. With these complications in mind, it is apparent that it may not be possible, or it may be very difficult, to mold some shapes. Large area parts with thin walls represent one class of parts that present difficulties. Because of the heat exchange between the flowing resin and the mold walls, the flow may freeze before the part is completely filled. Parts that have alternate sections with thick and then thin walls cause problems in flow and cooling that make them difficult to fill. In some cases the resins that have been selected for the end use requirement are too viscous to flow properly in a part, and this makes the manufacture difficult. Figure 10-1 shows some examples of cross sections that present molding problems. Thermo-setting parts that are transfer or injection molded combine thick and thin sections relatively easily since the hardening process is a chemical reaction. The injection molding of thermosets is an automatic transfer molding process. Annular shapes are best made by compression to gain best dimensional control and freedom from distortion. In the compression process, the molding compound is compressed and re-duced to the plastic state in the mold. During this process, portions of the material may lie in hard chunks in the mold while other por-tions are flowing rapidly with great force. Without proper preheating or mechanical plasticizing of the charge, portions of the product may be uncured and low in density. Transfer molding insures better proper-ties under average conditions. Impact materials that include long fiber fillers should be compression or plunger molded. Screw injec-tion will break up the fibers and produce weak parts.

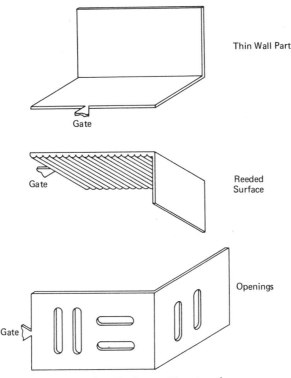

Fig. 10-1. Difficult to mold part sections.

Designing around these problems is the joint responsibility of the product and mold designers. For example, one way to handle the problem of thin, large area walls is by the inclusion of long ribs into the part in the direction of probable plastic flow. These ribs are not a functional requirement of the part but they act as auxiliary runners attached to the part to facilitate plastic flow in difficult to fill areas. In some instances the ribs may be used as a surface decoration like a corrugation or they may be on the concealed side of the part where they are stiffeners. Figure 10-2 shows some typical flow enhancement features of molded parts.

Another problem in molded parts is the existence of contiguous areas of thick and thin sections in the flow direction. In some cases this problem can be controlled by placing the gate in the thicker section of the part. Where there are several thick sections multiple gates are used. There is a limitation on this approach because the weld lines produced by the joining of the several plastic flows are a weak point

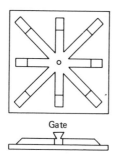

Gate

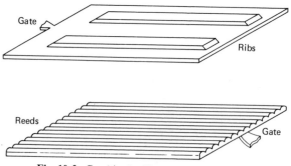

Fig. 10-2. Combined stiffeners and filling aids.

on the product. In some cases the use of a section that spans the thick and thin regions can be used to act as a built-in runner. This may be across the part or it may be a thickened edge or frame around the part. Still another approach would be to redesign the part for a more uniform wall thickness. The additional wall thickness required can be supplied on a mating part. Alternate approaches to the problem are shown in Fig. 10-3.

The preliminary discussion of the flow restriction problem was concerned with the simple and obvious necessity of filling the mold with material. There are a number of other consequences of restricted flow in molding which are less obvious and, generally, more signifi-cant to the performance of the part. Restricted flow parts cause high shear rates in the material as it fills the mold. This necessitates the use of higher injection pressures and usually the use of higher melt temperatures and higher melt index materials to fill the part. Higher melt index materials generally have lower impact and lower strength properties. The use of higher temperatures usually results in degrada-tion of the plastics properties by cracking the polymer to lower molecu-lar weight materials. Monomer or other low molecular weight break-down products are produced which drastically reduce the properties of the plastics material. The high shear rates encountered in molding also result in degradation of the molecular weight of the material

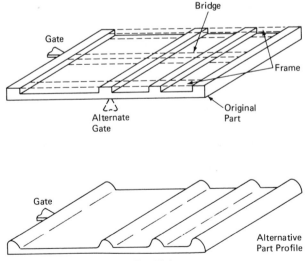

Fig. 10-3.

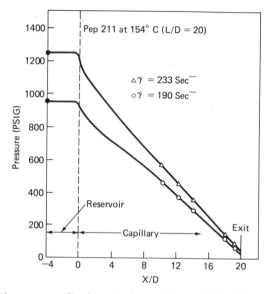

Fig. 10-4. Typical pressure profiles for polyethylene 211 at 154°C. L/D = 20. (From *A Study of Polymer Melt Flow Instability in Extrusion*, C. D. Han and R. R. Lamont. 29th S.P.E. Antec Book, pp. 553, 1971.)

just from the shear. The shear rate is directly dependent on the pressure drop in a channel and the pressure drop is a cubic function of the channel height. Figure 10-4 shows a curve which relates the shear rate to channel height for one polymer, giving some idea of how severely the material is affected by thin walls in the part.

The high shear rate produces two other effects that significantly affect part performance. The plastics molecules become aligned as a result of the high shear flow so that the material in the walls is highly oriented in the flow direction. This may be a desirable effect. For example, a restrictor bar is used in molding polyolefin parts to generate a living hinge effect by orienting the material. In some materials such as polyamides, the unidirectional orientation results in improved strength in both the flow direction and perpendicular to the flow direction. In most cases the effect is undesirable since the strength in the direction perpendicular to the flow direction is reduced and the part has a tendency to split along a flow line. In addition, the oriented materials have reduced elevated temperature properties in that the orientation tends to be relieved at a fairly low temperature and the part will distort as a result of the deorientation process. Figure 10-5

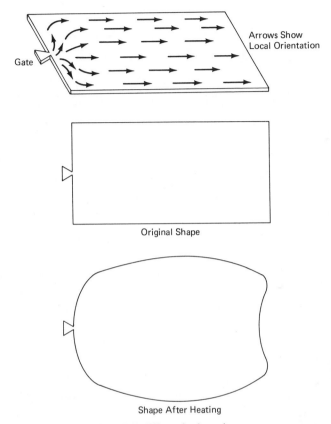

Fig. 10-5. Effect of orientation.

illustrates the way in which the walls of a part would be oriented as a result of high shear flow and the directions in which it would tend to distort as the temperature is raised to that at which the orientation starts to be relieved.

There is another condition that develops, particularly in parts with thin walls. This is a frozen-in stress, a condition that results from the filling process. This is illustrated in Fig. 10-6. The plastics material flowing along the walls of the mold is chilled by heat transfering to the cold mold walls and the material is essentially set. The material between the two chilled skins formed continues to flow and, as a result, it will stretch the chilled skins of plastics and subject them to tensile stresses. When the flow ceases, the skins of the part are in tension and the core material is in compression and the result is a

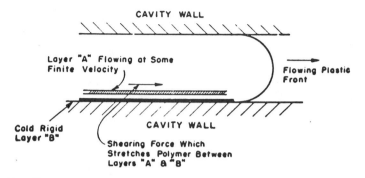

Fig. 10-6. Formation of frozen orientational strain. (From *Processing of Thermoplastics,* Bernhard, J. New York, Van Nostrand Reinhold, 1959.)

frozen-in stress condition. This stress level is added to any externally applied load so that a part with the frozen-in stress condition is subject to failure at reduced load levels. There are other conditions that result from the frozen-in stresses. In materials such as crystal polystyrene, which have low elongation to fracture and are in the glassy state at room temperature, a frequent result is crazing, i.e., the appearance of many fine microcracks across the material in a direction perpendicular to the stress direction. This result may not appear at first and may be triggered by exposure to either a mildly solvent liquid or vapor. Styrene parts dipped in kerosene will craze quickly in stressed areas. In any event, the crazing effect always leads to premature part failure. An annealing operation may minimize these stresses.

There is another result of frozen-in stresses that can be equally damaging to the part function and which affects materials that are not in the glassy state. This may affect an impact grade of material or a crystalline polymer even more drastically than a glassy material. The frozen-in stresses are real loads applied to the materials and when even slightly elevated temperatures are applied to the part, the stresses will cause the part to deform severely. This type of action is referred to as unmolding and it is a result of either the molded in stresses or the flow orientation relief or, commonly, a combination of both. It is a real disconcerting experience to see a molded part revert to a shapeless blob at temperatures as low as 110 to 120° in materials that have normally good performance at temperatures of 200° F and higher.

It should be pointed out that the problems we are discussing are interactions between the molding process and the part design. Poor control in the molding process can produce severe orientation and frozen strains in parts which are properly designed with respect to these problems. On the other hand, there are part designs used particularly with certain materials where it is impossible to avoid the frozen strain and orientation problems. It should be pointed out for use in the development of difficult-to-mold parts that the condition can be observed in transparent parts by the means of the photostress effect. Examination of a transparent molded part by polarized light will show the combined effect of the orientation and the frozen stresses. It is difficult to determine which effect is being observed since both have the same birefringence effect on the polarized light. The use of reflected polarized light from the surface of the part gives a somewhat different reading of the effect. This may be a way to separate the two effects which would be desirable since the result on the part performance is different for each condition. Much can be learned about thermoplastics product quality by "unmolding" the part as shown in Fig. 10-7.

It would obviously be desirable to make sample tooling and analyze the flow effects on a part which is likely to present a flow problem. In addition to the usual physical testing of the part, the use of photostress analysis techniques plus the exposure to selected solvents to check for stress crack characteristics would lead to changes in the product to minimize the effects of the molding on the part performance. There have been cases where piano keys with frozen-in stresses have unmolded from perspiration, leaving open flow lines.

The gate area on a molded part represents another processing problem in molded parts. Obviously the gate is the most severe restriction to flow since it is always desirable to have it as small as possible to reduce its visibility on the product. Because of the especially high shear rate on the material as it passes through the gate, the material is heated because a substantial part of the potential energy represented by the pressure on the material is converted to heat by friction. The effect on the material is drastic and, in the case of shear sensitive materials, there is substantial degradation in the molecular weight of the material as it passes through the gate. If the material is a filled one, such as a fiber glass material, the severe flow patterns generated at the gate will break up the reinforcing materials and convert fiber

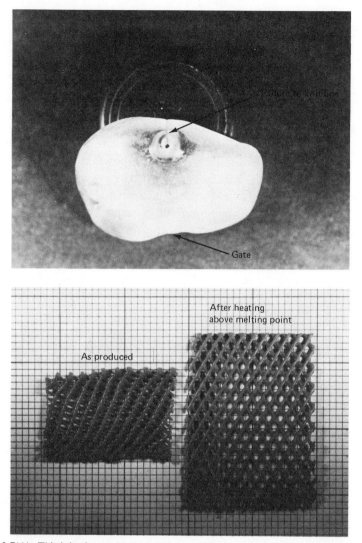

Fig. 10-7(A). This injection molded acrylic disc distorted as shown and note that the material did not knit above the pin hole—a failure-to-knit line occurs above the hole. This is a problem that must be anticipated when holes are located near an edge. **(B)** The extruded netting shows the effect of residual stretching stress and disorientation after heating. The extrusion process draws the material and orients it but due to the netting geometry this cannot be uniform. As a result, the netting shrinks by disorientation and distorts as a result of residual stresses.

to powder with a substantial loss in the reinforcing properties of the filler. The same thing happens when a screw plasticizer is used. There is not too much orientation that is permanently added to material as it passes through the gate since the continued flow produces turbulence that destroys the orientation. The last material to pass through, however, does retain its orientation and the gate area in a molded part is usually highly oriented and weak. In the case of jetting, which is a condition that results when the mold design has no immediate impediment to flow and the plastics is ejected into a relatively large open volume, the result is a patch of highly oriented material somewhere on the molded part near where the first material entered the mold. This jetted material becomes a weak point on the part and a surface blemish which is difficult to conceal. Jetting is controlled by changing the gate to direct the material to a nearby wall to slow the initial flow, by changing the size of the gate to reduce flow rate, by changing the shape of the gate, and by making adjustments in the fill rate on the part during molding. The design engineer's involvement in these areas is achieved by consultation with the mold designer and molder to indicate which options are available on the part that will not interfere with its function.

Weld or knit lines—the regions where two parts of the melt join while flowing into the mold—have been mentioned as a problem on molded parts. The quality of the weld depends on the temperature of the material at the weld point and the pressure present in the melt after flowing from the gate. The higher the temperature and pressure, the more complete the weld and the better the part performance and appearance. Bringing the material to the weld point at a higher temperature and pressure requires rapid filling of the mold cavity and, as we have discussed, this tends to produce flow orientation of the material and the possibility of induced frozen-in stresses which will also detract from part performance. In order to reach a reasonable compromise on these problems, the molder can contribute by running the mold at a higher temperature which will result in increased machine cycle time and higher production costs. The product designer can minimize the problem by increasing the wall thickness to permit easier flow, by the use of ribs to act as built in runners to improve and redirect the flow of material in the cavity, and by modification of the part design to shift and/or eliminate obstructions to flow. In some cases holes may be molded partially through the section to eliminate weld lines. Again the successful part design will be a compromise between the requirements of function, productibility, and cost. Mold makers will suggest mold designs that minimize weld lines.

Proper venting is essential for the successful molding of plastics parts. Air is eliminated by "breathing" a compression mold. Since venting may influence the product design, it is desirable to consider its requirement.

Differential expansion problems eliminate any possibility of molding gas-tight inserts in the plastics. Product designs that utilize two two different materials must include an allowance for differential expansion.

These are the primary process interactions that the design engineer must be aware of in order to determine process interference in part performance and design. Specific materials may introduce other problem areas as, for example, the problem of a level of crystallinity in a crystalline polymer part which exceeds the allowed level for stability of a molded part. The high crystallinity results from slow cooling caused by thick walls. Molded thermosets, whether molded by injection, transfer or compression, also have design restrictions imposed because of the chemical curing action that takes place during the molding and curing.

Certain specific problems occur with specific thermosets. For example, phenolic materials and others that evolve gaseous products during the cure have porosity problems caused by insufficient pressure applied to a particular area of the part. This lack of pressure in an injection or transfer molded part is caused by filling at too low a rate so that the pressure is not transferred from the gate to the remote portion of the cavity before the reaction causes the material to set up and block pressure transfer. In some cases this is a result of part design which has tortuous paths for the material to fill. It can be overcome by redesigning the part to increase the flow as suggested for the thermoplastic parts. Part of the problem can be corrected by changes in the grade of material, and in the design of the mold. Here, as in the other cases, the quality result will be produced by determining which factors are best changed in cooperation with the mold designer and molder.

For several basic reasons, the extrusion process does not have the large number of possible process product interactions that the preceding molding methods have. The process is a steady-state production operation and generally can be brought to a better condition of control. The operating pressures and shear rates in the extrusion process are considerably lower than they are in molding. Finally, as it exits the die, but not necessarily when it leaves the process, the material is in an essentially stress-free condition. Depending on the wall thickness of the material and the particular material, there is orientation of the polymer to a greater or lesser degree. Thin walls produce higher orientation in materials such as polypropylene, which is a highly crystalline polyolefin, and which orients much more than materials such as polyvinyl chloride and polystyrene. After the material leaves the die, it is usually drawn in size and passed through a set of jigs which reform the shape that is present at the die. The material is also cooled either by subjecting it to air flows, by immersing it in a water tank, or by subjecting the extrudate to a water spray. In some cases the material is drawn to a chilled metal mandrel by the use of vacuum applied to the mandrel. These draw-down, forming, and cooling procedures can and do introduce stresses in the part which affect the performance of the extruded materials.

One commonly encountered problem with extruded parts as a result of processing interaction, particularly with materials such as acrylics and vinyls which have "memory" characteristics, is that the part will shorten in the machine direction and thicken in the cross machine direction with the application of even low heat. This effect is analogous to the unmolding one we discussed in molded parts and results from the orientation produced by the drawdown process and frozen-in stresses produced because the drawdown was done at too low a temperature. Generally the die-induced orientation is not a major factor in this effect since it can be corrected usually by changes in the process. It should be pointed out that any orientation can cause this effect at high enough temperature, generally near the glass transition or crystalline melting point, and the designer should be aware of the fact that this is something to be considered in designing with extruded parts. The design engineer can exercise little control over this pull back condition except to be guided by the experience of the extrusion processor to indicate which materials are particularly susceptible to this problem and what the recommended wall thicknesses are to minimize the effect. In general, one of the best ways to improve the condition is to slow down the rate of extrusion. As a result, parts with this tendency to pull back will be more costly to produce.

Another problem with extruded parts is caused by distortion of the section by the effect of heat and other environmental conditions—exposure to water or chemical agents which tend to soften the material. These distortions are generally reversions of the profile back to the shape that it had exiting the die. Ths indicates that the post-die forming operations were done at a lower than desirable temperature which results in a molded-in stress. When the stress is relieved the part distorts. In some instances these stresses cannot be eliminated by process changes so that the part is inherently deficient in performance. The only way the design engineer can cope with this situation is to indicate to the extrusion processor what the anticipated operating conditions for the part are, so that the processor can design the tooling to minimize the potential for distortion. This involves (1) better control of the shape leaving the die achieved by more careful die design and correction so that a minimum of post-die shaping is required, (2) operating the line at a higher temperature when shaping jigs are used, (3) careful cooling of the extrudate, and finally by generally operating the process at lower rates to insure better control.

One general problem that exists with extruded shapes is in the control of dimensions. Because the production rate can affect the relative dimensions in an extruded part as well as the overall size, dimension control becomes an important economic factor. For economical production of extruded parts it is advisable for the product designer to indicate which dimensions are critical to the function of the part so that these are controlled, and to indicate the widest acceptable variance on other less critical dimensions. In this way the extrusion operator can adjust his process to the maximum speed consistent with the production of a usable product. Insistence on all dimensions as critical will result in the process being restricted to one narrow set of operating conditions, usually at a low production rate, with a high scrap rate and high product costs. In cases where the dimensions are critical, it may be that the extrusion process should be discarded in favor of molding or machining of the part.

Subject to the limitations indicated, control over extrusion process products is consistent enough to make a uniform product once the limitations are accommodated. Here, as in other processing, good communication between the processor and the design engineer will help make for a successful economical product.

Sheet forming processes, such as vacuum forming, do have effects on the product just as do the other processes that we have covered. The designer should be aware that these will affect the performance of his product and he should learn how to modify his design to minimize any deleterious effects. Probably the most serious problem encountered in formed sheet parts results from the fact that the materials are made from sheet at temperatures well below the melt softening point of the plastic, usually near the heat distortion temperature for the material. From our previous discussion on molded and extruded parts, it is apparent that forming under these conditions results in stretch orientation of the material and the production of frozen-in stresses. Since these conditions are unavoidable by the very nature of the sheet forming processes, the designer must accept the fact that the heat resistance of a sheet formed part and its resistance to other environmental stress factors will be lower than for a molded or extruded part. The objective of the designer should be to minimize the amount of stretching needed to make the part so that the overall performance will be as close to the molded part as possible.

Figure 9-25 illustrates the degree of stretching that occurs as a sheet is drawn down over several different male shapes. It also illustrates the variation in stretching that occurs in different portions of the sheet as the corners get sharper. There is a good correlation between the extent of stretching and the susceptibility of the part to damage; the degree to which it will occur will vary widely from one material to another. Table 10-1 lists the materials used for sheet forming in the order of sensitivity to property loss by the extent of the stretching. A material like rigid PVC or cellulose acetate propionate will

Table 10-1. Orientation Sensitivities of Plastics Materials.
(Materials are listed in decreasing order of sensitivity to molded-in orientation.)

Polystyrene
Polypropylene
Nylon
Rigid PVC
High impact polystyrene
Cellulose acetate butyrate
Cellulose propionate
High density polyethylene
Low density polyethylene
Flexible vinyl compounds

be much less likely to show damage when subjected to thermal or environmental stress than a material like polystyrene or polyethylene.

The nature of the damage to the part varies from one material to another. One temperature effect common to all materials results in unmolding where the part tends to revert to the original shape. The extent of this and the temperature at which it will occur will depend on the material, the forming process operating conditions, and the part design. With regard to the part design, the most stable thermoplastic parts are those with generous corner radii and smoothly blended surfaces with a minimum of sharp corners and re-entrant curves that will stretch the material excessively. Thermosetting materials are stable when properly molded. The best processing conditions will preheat the material to the highest possible temperature and then form the sheet as rapidly as possible. There are limitations on the process temperature because some materials such as polyethylene have a narrow range of forming temperatures while others such as PVC may be susceptible to thermal degradation. The speed of closure to the form is a function of machine condition and sheet thickness, with the thicker sheets being more difficult to move rapidly. The designer should indicate this to the sheet former when form stability at elevated temperatures is critical and the process must be tailored to improve this condition. He should also select a material which is intrinsically more stable and easily formed to minimize the possibility of unmolding. These factors must be considered in conjunction with the design of the part to minimize sheet stretching differentials in the part.

In addition to the unmolding problem, the effects of the stretching of the sheet materials result in two other impairments of the part. Highly stretched sections of the part are usually thin and highly oriented. The part has a decided tendency to split in these areas in a direction parallel to the stretching that took place. The designer should consider this and add material so that the part will perform adequately in use. The other effect of having a stretched area is a reduction in resistance to stress cracking. Crazing is a possibility in such areas in polystyrenes, and environmental stress cracking caused by solvent substances will occur in the stretched areas. This is a particularly important consideration in vacuum formed parts used for packaging food which frequently has some solvent action on the plastics.

With respect to the blow molding process, the type of situation where the material is stretched at temperatures below the melt temperature applies in a similar manner to that for formed sheet processing, only to a lesser degree. It is convenient to think of blow molding as a more generalized form of a plastics reforming process such as sheet forming. The comments on designing to minimize points of sharp stretching and excessive draw mentioned in sheet forming apply to blow molded parts as well. By parison control it is possible to minimize the wall thickness variation and the extent of stretching and stretch orientation. These are the province of the processor and, in this instance, the designer must rely on the blow molder to tell him what is possible and to select the processor that has the type of equipment and process that can make his part. The designer should be aware of the possible failure modes and allow for them in his design. There is little else the designer can do but select the best material and process to make his part. Blow molding materials must be selected after careful testing for permeability, etc.

The process interaction in cast plastics parts is mainly involved with the curing processes and with mold filling problems. Voids and porous sections are a frequent problem with cast parts because the mold filling is done at atmospheric pressure, or low pressure, and if the part has thin sections to fill, the flow may be a problem. Other than designing to avoid such difficult flow conditions and selecting a material with good flow characteristics that will perform properly, the designer must rely on the skills of the caster to make good parts. Frequently a cast part is selected because of the low tooling requirements and rapidity with which the part can be put into production. After the production level is increased and the requirements for better part quality are imposed, it may be desirable to change to pressure molded parts when the higher production level justifies the increased tooling costs.

The process effects on parts and the limitations they impose on part design and performance have been discussed for several of the more important production processes for plastics products. In many cases design modifications can substantially improve the producibility of the parts and reduce part cost with improved product quality. Understanding the effects of the process on the part is essential in making successful products. Specific information in this area is usually available from the processor who has experience with a wide variety of parts and knows the type of problems that have been encountered in the past. Because of the complexities of the materials and the effects of the processing on the materials, this is an area where predictability based on scientific data is limited and experience is desirable. Successful products will result from close cooperation between designer, tool designer, moldmaker, and processor to arrive at design compromises that make the part acceptable and producible at an economical price.

Chapter 11

Performance in Service and Environmental Exposure

The ultimate requirement of any part or structure is that it perform the function for which it is designed. With many materials the design life of the part is usually not as important as it is with plastics because of the creep characteristics of the plastics. In all cases the useful life is an important consideration whether the item is a pan for the kitchen or a bridge to handle traffic in a city. The people who use the designed product expect it to be properly engineered to perform satisfactorily in the intended environment for the indicated life, without endangering them or becoming functionally useless before the end of the predicted life. This, of course, implies that the user does not abuse the part and maintains it properly. It is the responsibility of the engineer to provide the user with sufficient information so that he can intelligently use the product and properly maintain it. No product can be guaranteed to perform properly if it is abused.

The responsibility of the producer and engineer was discussed earlier. This has recently become a matter of legal as well as moral responsibility. Improper design or inadequate safety instructions can lead to litigation in civil and criminal action. It is important for the design engineer to keep the ultimate user in mind when he designs a part or device. He must also make adequate allowance for human error and poor judgment to prevent malfunction and possible hazards to the user or to other property.

In this chapter we are concerned with how the designer discharges this responsibility. The effect of the use environment will be discussed as well as the effect that it can have on the performance of plastic parts. The design of suitable tests to determine if the part will perform in the end-use environment will be covered, as well as the cost of such tests. The test data will be reviewed to determine the potential hazards in accelerated testing made without useful physical principles to use as a guide. The method of designing around environmental effects will be discussed and safety factors for extreme operating conditions indicated.

Plastics parts are used most everywhere and it would be difficult to describe a typical use situation. One environment familiar to most people is the home and this environment has enough hazardous elements to serve as a useful starting point for our discussion. The kitchen is a particularly difficult environment. Some of the exposures that a plastics gadget or appliance might have in the kitchen are summarized in Table 11-1. The list includes a surprising range of physical and chemical environments that make the kitchen a torture chamber for plastics used in this area. If the stresses that the plastics parts encounter in use are included, it is apparent that kitchen items are rather difficult to design to resist the environment. For example, an egg beater has plastics gears that are exposed to chemical attack from food oils, acetic acid (vinegar), fruit juices, cleaning detergents, and

Table 11-1. Environmental Factors for Kitchen Appliances.

Heat
Cold
Water
Impact
High humidity at elevated temperature
Chemicals
    Soaps and detergents
    Oils, fats, and greases
    Fruit juices
    Phosphates (dishwasher cleaners)
    Fruit acids
    Caustic
Bleach, etc.
Biological exposures
Biological exposures
    Fungus
    Garbage
    Microorganisms
    Enzymes
    Vermin
Mechanical loading, static
Mechanical loading, dynamic
Water erosion effects

not infrequently, to heat when the beater is used on the stove, or laid down near a burner. The gear material is usually nylon or acetal resin and the beater is generally lubricated only by the food splashed on the gears by accident. The stress on the gears when beating a stiff mixture is frequently high since it is not unusual to use the beater exerting maximum force on the handle. The handle is exposed to the same environment although not to the same degree. The material selected for the gears and handle has good resistance to the environment as well as good mechanical properties or it would not represent one of the more successful applications for plastics.

There are other combination environments in the kitchen that are especially severe on plastics products. The first of these are the dishwasher and washing machine where the environment consists of heat and water exposure, both liquid and vapor. In addition, these washers use phosphates and detergents for cleaning. In the cleaning process the oils and grease from the food and clothes are heated to high temperature, increasing the solvent action on the plastics materials. Hard products may drop into the path of the impellers. The second is the refrigerator/freezer where the low temperature and high humidity can severely affect plastics. Plastics in the refrigerator are used in the proximity of food and the oils and essences in the foods are also damaging to some plastics.

After the preliminary design of a product to be used in one of these environments, it is essential to determine the effect of the environment on the plastics part. In the case of the dishwasher, it is important to see that the maximum safe use temperature is not exceeded. Exposure to elevated temperatures and humidity will accelerate any tendency to unmold, especially in plastics which absorb significant amounts of water. Water acts as a plasticizer and lowers the softening temperature of the part. Detergents, oils, and phosphates present in the dishwasher weaken the part. The materials tend to soften the plastics or to increase the possibility of stress cracking. Another problem is that expansion and contraction of the part during a cycle may cause distortion of the part.

To design around these problems it is necessary to do several things. First, the material selected for the part should be examined for chemical resistance to the environment and for adequate temperature resistance. Some of these data are available but, it should be pointed out that a combination of several environmental factors such as heat and

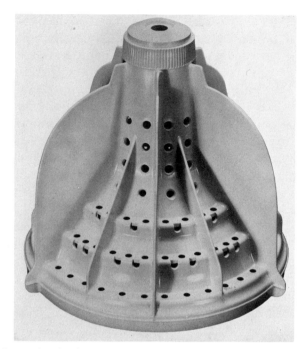

Fig. 11-1. Polypropylene is widely used for housewares and appliances since it will withstand hot water, detergents, mechanical abuse and is low in cost. This injection-molded washing machine agitator is a typical heavy-duty application served well by polypropylene. (*Courtesy Hercules, Inc.*)

chemicals can have a synergistic effect that produces a result much more drastic than the overall individual attack possibilities. After an appropriate material has been selected, it is essential to devise a test that will simulate the use conditions and do some accelerated testing to determine the life of the part under actual use. It is important that the test not be misleading and that the exaggeration of a single factor to accelerate the test does not distort it to give results that either eliminate a useful possibility or show a potential failure to be useful.

An example will be given to bring the factors into perspective. Suppose we are concerned with designing the agitator arm that moves the water around in the dishwasher. This unit introduces the water into the washing space and continues to spray it around. The part rotates rapidly and is subjected to internal water pressure as well as to centrifugal and axial driving forces. The environmental factors are—heat, stress, chemicals, water, and high humidity. The maximum temperature for the dishwasher is about 180°F although the temperature can rise above this during the drying cycle when parts are exposed to radiant heat from the calrod heater. The first step is to select a material with a heat distortion temperature well over 200°F that is resistant to moisture. The material should be fairly strong and stiff to resist the pressure and the centrifugal forces. Using the accelerated temperature apparent modulus data, the part is designed to withstand the stresses for the anticipated design life which will be about 10 to 15 years for this type of appliance. The selected material should be checked for odor and general chemical resistance to the potential exposure. Some of the materials that could be used for the part are glass-filled nylon, polypropylene, glass-filled polyester, mineral-filled phenolic (used for 30 years), or a glass-filled ABS material. In general, these materials have the required temperature and chemical resistance.

The next step is to see what type of failure might occur and what the consequences of the failure might be. The unit could fail by softening and be distorted so that it would not function properly. It could jam the rotor and cause a motor burnout. The internal pressure and heat could cause the part to distend while rotating, jam, and fly apart with the fragments damaging the contents of the machine. The combination of the stress and chemical action might cause the part to fracture. Pieces of it might fly around and damage the machine— break the calrod heater or the contents of the machine. It is also conceivable that some of the fragments could exit the machine and strike someone. Machine damage may cause leaks or overfilling and flooding with possible electric hazards.

With a list of potential failures the designer examines the part design to determine how to change the geometry to either strengthen the part or to reduce the possibility of failure. The material should be evaluated from existing data to determine the possibility that combinations of environmental factors will not lead to premature failure. For example, the polyester-based material has good resistance to chemicals and water but the combination with high temperature and stress may be damaging. Nylon is very resistant to hydrolysis even at elevated temperature and stress. However, the combination of detergents and alkaline phosphates may cause excessive water absorption which can soften the material and make it susceptible to failure by softening. The ABS material has excellent resistance to heat and water but detergents and food oils may cause it to become brittle. In each case the combination environment can have effects much more severe than each one by itself. After evaluating the data the next step is to devise a set of tests that will show the actual performance of the part. Since it is impractical to do a total life test, an accelerated test is essential.

The various factors that affect the part cannot be exaggerated for the accelerated testing without considering also the direct effect of accelerated testing. Obviously, the test should be done at maximum humidity and include water immersion. It would be difficult to accentuate this beyond the conditions that actually exist in the machine. The chemical environment can be simulated by using the actual materials at somewhat increased concentrations. Here caution is in order since many plastics materials are very resistant to low concentrations of materials and become increasingly sensitive to higher concentrations. Usually the chemical resistance data include tests at several different concentrations so that it is possible to determine how far to go with increasing the concentrations to accelerate the testing. Temperature is one factor that is usually increased in accelerated testing. The temperature should be increased with care. A plot of the modulus vs temperature for the material should be studied as shown in Fig. 6-2. Some materials soften gradually with temperature over the range that will be used. Others may undergo a drastic change with a relatively small temperature change. Among the examples selected, the ABS material should be exposed carefully to increased temperature while nylon and polyester soften gradually in this range.

Increasing the load level is another way of accelerating the test. Here, again, it is important to do this cautiously and base it on the performance curves for the materials. The creep of the plastics is sensitive to the stress level and the range under consideration may be such that a relatively small increase in stress level will result in a large increase in creep level. In addition, since the part is also dynamically loaded, the increased stress level may lead to severe heat build up due to increased size of the hysteresis loop at increased stress levels and this could lead to catastrophic failure of the part. In any event, the data for the material should be carefully examined to see that the acceleration method used does not lead to erroneous results because any critical level factor is exceeded.

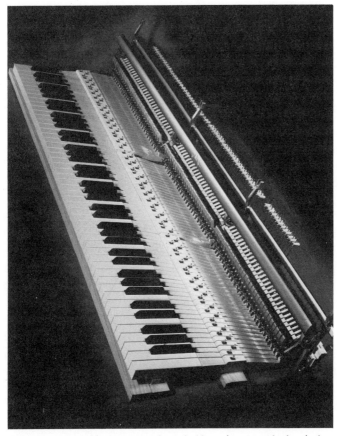

Fig. 11-2. Piano keys are subject to severe household, environmental, chemical, perspiration and scratching. (*Courtesy Tech Art Plastics Co.*)

Another tactic to be used in accelerated testing is to inspect the part for degradation short of total failure. Excessive distortion of the part will indicate the probable onset of failure long before the part actually becomes nonfunctional. Effects of the chemical factors can be indicated by surface damage to the part, by reduction in the modulus of elasticity, or by loss of tensile and impact strength and weight changes. Some of the parts can be removed from the test at various intervals and examined to see the rate of deterioration of key properties. From this it is possible to determine the rate of degradation and the probable life expectancy of the part.

It is evident that testing is not a simple thing. Basic test data on the materials cannot be relied on except as a generalized guide and as a screening method. Tests of similar materials in similar environments are much more useful and there are several sources where such data can be obtained.* Lacking better guidance, the engineer is forced to evaluate the basic nature of the materials and, extrapolating the available data, devise a test that will indicate the probable survival time of the product in the environment in the minimum time with reasonable expenditure of effort and cost.

The refrigerator environment represents a somewhat different problem. Existence of the static environmental factors such as the cold and relatively high humidity and the effects on the material properties with time must be evaluated just as in the case of the dishwasher. Materials become brittle at low temperature and the absorption of water over a period of time can affect the performance of plastics parts. It is necessary to evaluate the degradation of the materials with time and determine effective lifetimes for the parts. One problem that exists with this environment that complicates the problem is the use factor. For example, it is possible to use a material with adequate impact properties for door liners or for shelves which, with normal handling, will last indefinitely. However, the possibility of sharp impact at low temperatures where the impact strength is low does exist. If the door is closed on an object, or if a container of food drops on to a shelf or drops against a wall or support, the part can be damaged. Another type of possible in-use hazard involves breakage of a container and the spillage of a particularly potent product such as a fruit juice or a product containing a large amount of oil. The butter test for transmission of odor and taste to butter is commonly used. Still another in-use hazard is a diligent cleaning person who will attack the interior of the box with strong chemical cleaners and bleach to keep it smelling clean.

Normally the materials used in the interior of a freezer or refrigerator are not selected to take this type of abuse. The main reason such materials as polystyrene and impact styrene are used is that they are relatively inexpensive and do not readily absorb food odors. Economics is a major factor in material selection and the use of higher cost materials with good resistance to abuse does not seem to be justified. Here the responsibility of giving proper instructions in use becomes apparent. Failure of a part in the interior of a refrigerator or freezer generally does not create a high risk situation. In most

---

*Plastec & General Electric Information Services.

cases the failure does not result in a malfunction of the freezing ability of the machine. Here the major problems are the possibility of fractured plastics getting into the food or sharp pieces of broken plastics that may cause injury in handling. With the relatively low risk situation, apparently the design decision in this area is toward economical materials. What is generally lacking is proper instruction in use to prevent failure caused by breakage of the plastics from any of the above causes. The mere fact that plastics can crack and be potential hazards is no reason not to use them. They typical refrigerator contains jars made of glass as well as bottles of all descriptions and people are accustomed to handling these. Being properly cautioned that the materials are susceptible to damage by impact at low temperature and sensitive to chemical attack from food and cleaners would prevent most of the damage. Then, in the event of damage it will not be considered a "poor" product. It would be better if the machine were abuse resistant but not if this were to increase the cost to the point where it is not an economical product. The designer, as part of his responsibility, must learn to exercise judgment as to where usage instructions are necessary. This will prevent the assumption that the machine is a juggernaut that can resist all abuse.

Other household environmental problems are presented by the piano keyboard (Fig. 11-2) and the furnace humidifier (Fig. 11-3). Piano keys must be resistant to household chemicals, cleaners, foods, and perspiration. Organs played in night clubs often have melamine keys for improved cigarette resistance. Piano sharps require a scratch resistant material since the player may strike the keys with his fingernails; organ sharps can be thermoplastic because the player rolls over the keys. Styrene acrylonitrile (SAN) is commonly used for natural keys and for telephone handsets that encounter the same problems. Bottle caps often crack (Fig. 11-4) when fastened extra tight and craze from exposure to environmental products, conditions, and lost liners. Products designed for laboratory or other use must be tested for their resistance to environmental chemicals as illustrated in Fig. 11-5.

Plastics are exposed to a wide range of other conditions that require consideration of the effects of the environment and of the in-use handling to make the product operable over the anticipated design life. One of these of extreme importance is outdoor exposure—the effects of weather, exposure to sunlight, wind and wind-driven water and solids, the effect of chemicals in the soil and air, and the effects

Fig. 11-3. The cabinet, access door, motor assembly and drum assembly of Lennox's central humidifier is molded from polyphenylene oxide (PPO) to solve corrosion problems that exist with metal components when exposed to the various water conditions across the country. The grade selected, is rated SE-1 according to U.L. Bulletin #94 on flammability. This and the material's high heat resistance of 265° F enabled the units to be certified by the Air Conditioning & Refrigeration Institute as passing all requirements of their standard #610 which measures capacity and ability to withstand a 200-250° F temperature. (*Courtesy Lennox Furnace Co., Marshalltown, Iowa*)

of bacteria in the soil. A plastics material used for a fence or garden hose or for house siding will be affected to varying degrees by the general outdoor environment. Plastics used for such applications are generally specially formulated for resistance to ultraviolet light, microorganisms attack, and the effects of humidity. Some materials such as the fluorocarbons, acrylics, and rigid vinyls have good inherent resistance to the outdoor environment because of their chemical structure. These properties are enhanced in the vinyls and acrylics by the use of additives to reduce the effects of UV light and microorganisms. The design of parts for these environments should start with materials that have had a successful history of use out-of-doors because the range of conditions for outdoor exposure is so wide that effective testing is an expensive and time-consuming process. Materials are constantly under evaluation for outdoor use since some of the most important applications are in outdoor products. Generally

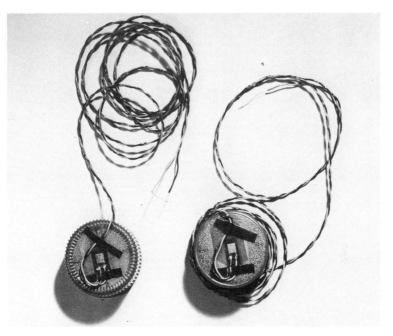

Fig. 11-4A. Resistance strain gages attached to the tops of plastic bottle caps are used to study stress increase as the cap is tightened. (*Courtesy Dow Chemical Co.*)

Fig. 11-4B. Plastic bottle caps which were coated with a brittle lacquer prior to over-tightening by various degrees. After the coating cracks, a die is allowed to stain the cracked areas. This allows the analysis of direction of principal stresses.

Fig. 11-5. Milking machine parts. The thermoplastic, Polysulfone, was selected by the Zero Manufacturing Company because of its superior resistance to chlorinated cleaning solutions, light weight, and toughness.

a large amount of data are available on those materials that the manufacturer feels will perform under outdoor exposure.

It is important to analyze the type of deterioration that takes place in the materials out-of-doors and to evaluate the performance requirements of the products. One case in point is the use of acrylic materials for signs and skylights. Here, the risk factors are significant and deterioration of the material can lead to failure. Signs are frequently mounted overhead and failure can result in parts falling on passers-by with possible injuries. This condition can be more serious if failure occurs in a high wind and fragments are carried about with force. The hazards of skylights are similar. There is an additional problem of weather exposure to the contents of the structure resulting from the failure of skylights, thus the effect of weather on the material is important. In the case of acrylics outdoor exposure gradually embrittles the material and renders it more likely to failure by impact. To minimize the failure it is desirable to mount the parts so that fracture of the part is lessened by cushioning mounts and by having something to catch any broken parts. Because of the danger involved new acrylic materials have increased impact strength which is retained over longer periods of time because of better formulation. The acrylic materials also indicate the possible incidence of failure because the surface quality of the material is reduced as the part reaches the age where failure is likely. In this instance the requirement is to alert the user that this is a possible hazard and to inspect for damage and replace the part before it deteriorates to the point of presenting a problem. Microscopic inspection is a useful tool for finding change.

Vinyls have been used extensively as building materials for siding, roof barrier materials, window framing, gutters and downspouts, and other similar products. The use of plastics for such applications introduces another type of consideration for the design engineer. Traditional building materials have what is considered to be an indefinite useful life whether or not such a concept is justified. Masonry structures and properly maintained wood structures can last for centuries and this type of performance is expected in building materials. Plastics used out-of-doors can not usually be expected to have such an endurance factor. Even properly formulated, vinyl compounds with stable pigments are difficult to guarantee for periods longer than 10 years and such assurances are based on long-term in-use testing and evaluation. The vinyl materials generally show surface attack and the incidence of embrittlement after 5 years, and the possibility of failure increases after this. There are several ways in which this can be dealt with. For example, siding can be thought of in the same way as a paint job and the material can be periodically replaced. Materials can be developed to coat the vinyl and extend its life so that it may be considered to be indefinite. Such coatings based on the fluorocarbon resins are, in fact, used. Maintenance coatings can be developed which are, in effect, paints to coat the vinyl periodically and thus restore the surface and to some extent the bulk of the material. Such materials are under development and are used to some extent. In any event, this detracts somewhat from the use of the plastics product but it does make the required durability for building products a more realistic possibility. It is necessary for the designer to consider the needs of the market and to satisfy the requirements for permanence rather than risk disastrous failures of the materials with serious consequences to both the user and the manufacturer.

There is a more general area of application for plastics parts in extensive use that requires other considerations. In using plastics for applications that range from one of pipes carrying chemicals to severely loaded structural parts such as helicopter rotors, the risk involved is usually much higher than it is in the consumer applications covered above. Failure of an acid carrying pipeline or a helicopter rotor implies serious problems of life hazard and destruction of property on a large scale. Here the material evaluations must be done more carefully and the testing must be more sophisticated and extensive. A systematic approach is essential for the evaluation of a plastic material and this must be backed up by a quality control system to insure that the actual parts are made to perform properly. The steps are:

1. Material selection to meet the requirements of the application, followed by materials testing to indicate the performance over the range of operating conditions expected.
2. Construction of test samples that can be subjected to actual exposure to the end-use environments which are tested to destruction to determine the possible modes of failure and the conditions that cause the failure.
3. Design of the part to meet the anticipated environmental stresses as well as the functional requirements. The part must have a designed-in means for reducing the possibility of failure plus either a fail-safe mode or an indication of incipient failure that can be monitored.
4. Extensive testing of the part in simulated service as well as accelerated testing. The testing program should include a method of evaluating the effects of fabrication procedures on the part performance and methods to inspect for compliance with the proper procedures. This may include destructive testing, x-ray inspection, ultrasonic testing, and a wide range of other sophisticated testing procedures.
5. Introduction of the part into limited service with constant monitoring of its performance. As the experience factor on the part performance increases its use can be extended. The continuing testing of the parts under the procedure in step 4 is essential so that if degradation in the part performance is seen, it will show up in the tests before it becomes a serious problem in the end use.
6. Continuous monitoring in use over a period of several years to insure continued performance and a replacement program that will take parts out of service after a conservative use-life to preclude possible failure.

This type of cautions approach is essential in high risk applications since the plastics materials have had limited experience levels and only extended experience can be used as the basis for such applications. It is true that such procedures are not always followed, and this has led to serious problems in a number of instances.

This chapter has been concerned with the problems of the product in use under the types of environments and potential abuse that may be encountered, and how the design engineer can prevent premature product failure which is one of his major responsibilities. Determination of the hazard potential and designing around it is one element of the solution. The use of carefully designed accelerated and continuous testing is another element. The instruction in use and proper maintenance procedure is the third element. By exercising judgment as to the appropriate combination of these elements consistent with the economic factors involved, the designer can have a product that will perform for its projected design life with a minimum of hazard to the user and with a high degree of satisfaction in use. The application area is so large that no cookbook advice would be useful. Each type of application must have the potential failure cost evaluated, the economics calculated, the degree of assurance of performance set, the extent of required testing determined, and the product designed and evaluated to meet the criteria established. There are sources of data available on which to base the approaches to the requirements, but the final combination of factors must be determined by the designer.

Chapter 12

Design Procedure for Plastics Parts: Function, Material, Geometry, Test

This chapter will outline the procedure for the design of plastics parts and give the methods of sizing, material selection, cost consideration, design life, performance evaluation, and testing. The monitoring of quality will be indicated to insure performance. All the basic information and approaches have been covered, thus this chapter is a synthesis of previous discussions.

A part under consideration must have some utility; it must fulfill a need which may be aesthetic or functional and, generally, both. Since we are concerned with engineering, we will not consider the aesthetics except insofar as the design and manufacture do not detract from the appearance quality. To proceed with the design we must know what function the part is to perform. We also need to know the context or surroundings of the part to determine what effect they will have on its function. A careful definition of the function will simplify the design and permit the widest latitude of alternatives possible in the design without compromising the function of the part. The comment made earlier concerning the shelf is a good example. The function served is a support which can hold several objects in a desired location for storage or for display. Unless a more specific function is defined, the shelf can take on a wide range of possible shapes, structures, and materials. It would be necessary to define the shelf as a bookshelf, for example, compatible with the environment found in a library, suitable for holding five or more books per foot of shelf, and for constant reference use rather than for storage. This will sufficiently define the function so that a design can be started. This limitation defines the environment, sets the load level and the type of loading situation, and gives some idea of the shape requirements, as well as the possible aesthetics of the unit. It still permits a wide range of design choices as to material and structure and shape but they would be limited to those normally used in a library environment. The more accurately and completely the function is defined, the more restricted are the design possibilities and the more detailed the specifications for the function.

Size is the next factor to be considered. A part has to fit its function within the confines of the space in which it is used. Continuing with the example of the shelf, it is obvious that we must know the length of the shelf, either by deciding how many books it will hold or by stating the size of the supporting rack that it will be used with. The size can then be decided either by burden or by space restrictions. In most cases, one or the other of these considerations will apply and in a typical case both may apply to some extent. In the example given, the width of the shelf would be determined by the width of a book, which ranges from 6 to 11 inches. The typical bookshelf is supported at 36-inch intervals so that the shelf would be this wide to fit a typical book rack. The shelf will hold about 5 books per foot with an average weight of about 2 pounds to make a maximum load on the shelf of 30 pounds. If the shelf were completely filled, it could be considered a distributed load, or it can be considered as a set of discreet loads.

The type of shelf design is the next consideration. The shelf can be a solid plate of plastic material, an inverted panlike structure with reinforcing ribs, a sandwich-type structure with two skins and an expanded core, or even a lattice type sheet which has a series of openings. The several possibilities are shown in Fig. 12-1. The choice between these is dictated by a number of factors. One is appearance or aesthetics. The lattice-type shelf is functionally as good as the others, but it may not look appropriate for a book shelf in the context of a library. A second consideration is a combination of physical requirements and appearance. A simple plastics beam that will function adequately in terms of strength and stiffness may be rather thin. A shelf of this type can look flimsy even if it is functional. This impression is useful to the designer since the solid plate is probably an uneconomical use of material. The design should look like a wood shelf since this is the context in which it is to be used. To produce the desired thickness appearance either a lipped pan with internal reinforcement can be used or, alternatively, a sandwich-type structure with two skins and a separator core. In either case the displacement of the material from the plane of bending will improve the

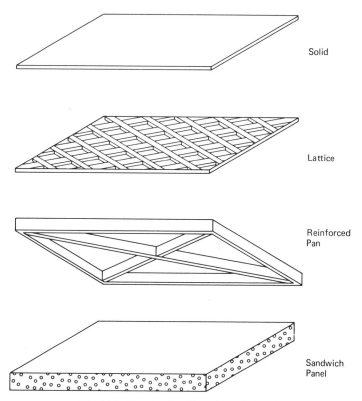

Solid

Lattice

Reinforced
Pan

Sandwich
Panel

Fig. 12-1. Structural sections for a chairseat.

stiffness efficiency of the part. The appropriate procedure is to leave both as possibilities and do some trial designs.

The next step in the design procedure is to select the materials. The considerations are the physical properties, tensile and compressive strength, impact properties, temperature resistance, differential expansion (Fig. 12-2), environmental resistance, stiffness, and the dynamic properties. In this example, the only factor of major concern is the long-term stiffness since this is a statically loaded part with minimum heat and environmental exposure. While some degree of impact strength is desirable to take occasional abuse, the part is not really subjected to any significant impacts.

Using several materials such as polypropylene, glass-filled polystyrene, and styrene molded structural foam which is a natural sandwich panel material, the design procedure follows to determine the

Fig. 12-2. This phenolic part has a steel insert extending 90% of its length. When the pot was heated to cooking temperature, differential expansion broke the handle as shown. The problem was solved by having only a short section of the insert extending into the handle.

deflection and stress limitations of the material in each of the several designs. There are two criteria to use as the basis for evaluation. The design life of the part is determined by deciding what the part will tolerate in deflection and still be useful. This is combined with the cost of the part to arrive at the cost effectiveness value that the part must meet. For example, we can say that if the part costs N the life must be A months, if it costs M it must last B months, and if it costs P, it must last C months. This can be presented as a table or it can be graphed as the criteria range that the part must meet. Using the several materials selected and the basic design possibilities, parts should be designed to meet the criteria as far as deflection is concerned, and the cost of manufacture estimated. This is best done with the assistance of the processor who will make the part. In addition to the material cost and the production labor costs, the amortization costs of the tools should be included. The various designs and costs can be tabulated and the ones that are the most economical can be determined.

At this point, it may become evident that the design life can be long and the cost of increasing the design life small, or, alternatively, it may be that the cost of small imcrements in the design life are quite costly. In the latter case, the design life should be limited to the acceptable minimum. A value judgment must be made as to the product quality requirements and the final design made to meet these requirements. By leaving the options open until this conclusion is reached, the decision as to what is best in terms of product can be based on more than a single valued solution, with the probable result that a more economical and practical product will result. The plastic parts shown in Fig. 12-3 illustrate good product design and material selection.

The next step is the performance evaluation of the part. In order to do this the part must be completely designed as to the dimensions

Fig. 12-3. Used to connect signal cables in oil exploration, this connector, which is made by Hughes Connecting Devices of General Electric's VALOX® thermoplastic resin, is designed to withstand all types of field abuse: tension loads of up to 3000 pounds, resistance to hydrocarbons, salt water and acids.

necessay to fit any surrounding parts, as well as to the necessary cross-section thickness for strength and stiffness. The material, color and manufacturing process must be selected. The part must include in the design the necessary features to make for proper processability and whatever design features are necessary to improve the performance in the use environment. The latter may include reinforced areas, coating or plating, inserts, etc.

The next step is to make sample tooling and sample parts for the test. Samples made by machining or other simplified modelmaking techniques do not have the same properties as the part made by molding or extrusion or whatever process is to be used. A part made this way is a sample rather than a testable prototype. Simplified prototypes may reduce trial mold cost and produce adequate test data in some cases. Its main value is appearance and feel to determine whether the aesthetics are correct. Any testing has to be done with considerable reservation and caution. If prototype tooling is not made, and frequently it is not, then production tooling must be made to provide samples for evaluation testing. This is justified if the part does not represent a substantial departure from previously made units whose performance is known in similar applications. The risk that the tools may have to be scrapped or drastically altered as a result of the testing is not high and is justified. The other reason that a production tool is made with no prototype tooling is because of the lack of lead time. Here the risk is usually not justified and the shelves of processors are littered with tools that were the result of bad guesses made under severe time pressure. By proper organization of time and tool design, the sample tooling can be made to fit into the normal tooling cycle with a minimum of added time and expense. Sometimes the prototype tooling can be made part of an existing master chase or holder or it can be made of soft materials such as epoxy or kirksite which can be worked rapidly. In any event, the information obtained in the prototype tooling stage can have a major effect on the final design and tooling to give a successful and economical part while prejudgments may result in a poor compromise as a result of drastically reworked tooling.

After obtaining the prototype parts, tests must be made to determine the utility. Generally these include a short time destructive test to determine the strength of the part and to check out the basic design. Another test that is done is to use the part in the projected environment with stress levels increased in a rational manner to make for an accelerated life test. Other tests may include: consumer acceptance tests to determine what instructions in proper use are required, tests for potential safety hazards, electrical tests, self-extinguishing tests, and any others that the part requires. In the case of high risk parts, the test program is continued even after the part enters service.

The remainder of the test program requires the generation of quality control monitoring tests. For example, in the case of an injection molded part such as the library shelf, the quality control monitoring might include critical dimension checks and an oven test to see if the part tended to unmold at low temperature, indicating improper molding conditions. A solvent test for monitoring proper process conditions might be included to check for stress crack resistance.

Table 12-1 summarizes the steps in the design of a plastic part. It indicates the different types of options that can be used at different stages in the design sequence. It also indicates the areas of potential high risk failure and the alternative approaches to use under these conditions.

Table 12-1. Steps in Plastics Product Design.

1. Define the function of the part with life requirements.
2. State space and load limitations of the part.
3. Define all of the environmental stresses that the part will be exposed to in its intended function.
4. Select several materials that appear to meet the required environmental requirements and strength characteristics.
5. Do several trial designs using different materials and geometries to perform the required function.
6. Evaluate the trial designs on a cost effectiveness basis. Determine several levels of performance and the specific costs associated with each to the extent that it can be done with available data.
7. Determine the appropriate manufacturing process for each design.
8. Based on the preliminary evaluation select the best apparent choices and do a detailed design of the part.
9. Based on the detailed design select the probable part design, material, and process.
10. Make model if necessary to test the effectiveness of the part.
11. Build prototype tooling.
12. Make prototype parts and test parts to determine if they meet the required function.
13. Redesign the part if necessary based on the prototype testing.
14. Retest.
15. Make field tests.
16. Add instructions for use.

The last step in the design cycle is the end-use testing, or field trials. This can be done using selected individuals or applications which are closely monitored or controlled. This information will indicate whether or not the part performed as the designer anticipated because of some unexpected situation. For example, the library shelves are often cleaned with lemon oil which disintegrates the polystyrene. Some of the product can be test marketed to uncontrolled users and their reactions sampled with a standard response form generated by the advertising people. Sample responses from a larger group of people frequently produces what might be referred to as the "idiot response." It is unbelievable what some users can do to a product simply because they did not understand its limitations. Loading an injection molded styrene Parsons table with two boxes of cans weighing 75 pounds is an example that the author has seen. An analysis of the impossible situation data is useful in finalizing the design procedure that is the instruction for use. Despite the fact that plastic materials and plastic products have been widely used for over 50 years, most people do not recognize their limitations and their differences from other materials. This may result in part from the early approach for selling the materials as substitutes and from a lack of discrimination which seems to exist in the plastics field. Few people know the range of materials covered by the term plastics. Many are surprised by the extent to which plastics are used in difficult applications.

The results of the field testing must define the basis for labeling and the instructions to be used with the product. Unless there has been a serious misjudgment on the part of the product development people, the field tests do not lead to redesign of the part. There may be the need for a more durable unit for the serious abuse situations, as well as the need for proper instructions, but the main need is for the labeling and instructions. The design engineer will supply the necessary data for the do's and don'ts and the instructions will at least minimize liability on the part of the producer for failure due to abuse. The most obvious case in which this is not true is when the abuse may occur easily and result in a personal hazard to the user. In this case, loss of the instructions is not an adequate defense against responsibility by the user. In all cases, the part must be designed to prevent danger to the user. Potential failures that can cause personal injury must be avoided in all product designs.

Chapter 13

Design of Plastics Structural Parts for Static Loads

This chapter is concerned with the detailed procedure for the design of plastics parts to take static loads. In order to make the analysis significant to real problems, we will work with a situation that involves significant loads and important structural requirements. In order to generalize the approach we will work with a structural problem common to a number of different structures and which will show how the different structural requirements will affect the engineering choices to be made. The particular structural section that will illustrate the sample design problem will be a roof section which may be used for anything from a work shed, to a house, to a vehicle, or even to a simple lean-to weather shelter.

The analysis begins with a definition of the function that a roof performs. A roof is the overhead part of a structure intended to protect the occupants and/or contents of the structure from the outside environment—that is, from rain, snow, wind, sun, falling objects, and the other elements that make up the outside or surrounding environment. In order to perform this function the roof must be capable of supporting its own weight and the weight of snow or any other possible accumulations on the roof. It must be resistant to wind loads which are quite severe in some regions. The roof must also support loads imposed by people walking on it, usually for maintenance. In some instances the roof may double as a deck and the traffic may be constant. The roof must be able to shed water which falls on it, although it need not be waterproof in the sense of being a waterproof membrane structure. The roof surface is exposed to sun, wind and driven debris and must be resistant to erosion by the action of sunlight and the abrasive action of wind driven debris. In most cases the roof is insulated thermally to prevent heat loss in cold weather and heat input during warm weather. Obviously, not all roof require-

Fig. 13-1. Plastics house of the future. (*Courtesy M.I.T.*)

ments apply to all roof situations, but most of them do. An all-plastics house is depicted in Fig. 13-1.

The major load applied to a roof is the static load of the roof structure itself. Since roofs come in a wide variety of types the self load will depend on the basic roof design. Figure 13-2 shows a number of typical roof structures in which plastics are frequently used. The simplest is the corrugated fiber glass panel structure shown in Fig. 13-3. This type of structural element is widely used for roofs on industrial buildings to admit daylight, for porch and patio roofs, for shelter roofs such as those used at bus stops, and for a variety of similar applications. Variations of this simple roof are used for roof sections on transportation vehicles such as buses and trains. Since this section is one of the easier ones on which to describe loading conditions, it will be used to illustrate the design procedure. Other roof sections such as the domes, arches, geodesics, and paraboloids involve

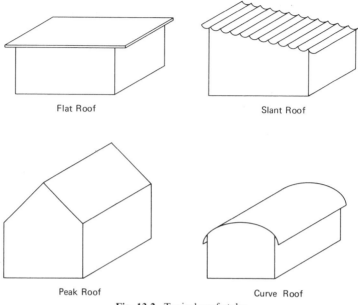

Flat Roof   Slant Roof

Peak Roof   Curve Roof

Fig. 13-2. Typical roof styles.

complicated stress analysis and the results would not be particularly useful in a general analysis of static structure.

The corrugated materials are available in sheets which vary from 4 feet × 8 feet to as large as 10 feet × 20 feet. A typical material is .100 inch thick with 2 inch corrugations, and a corrugation depth of 1 inch. The plastics material from which they are made is a polyester glass material using glass mat as the reinforcement and a weather-resistant polyester resin. In general, the sheet material is nailed or screwed to wooden supports at proper intervals. In some cases the roof section is made in one piece with spars of polyester glass material molded into the part to provide the stiffening support needed. In this case the only requirement for installation involves anchoring the edge of the section to the structure.

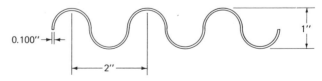

Fig. 13-3. Typical corrugated plastics section.

This type of design problem is somewhat different from the previous examples in that the unit is made from standardized sections which have specific physical properties and are available in only a limited number of thicknesses and configurations. The design problem now consists of trying the available materials in the structure with the supports that can be used and then determining if the material will work. The self load is easily determined from the weight of the materials. The snow load is a design value available from experience obtained in the area where the structure is to be used. Similarly, the maximum wind load and people load can be determined from experience factors which are generally known. The problem is worked out in Table 13-1 using several different sheet types and different support spacings in an environment which would be typical of a city in the midwestern part of the United States. The indicated solution is that the material selected will take the required loads without severe sagging for a 15-year period with little danger that the structure will collapse due to excessive stress on the material. If a standard material had not been suitable, it would have been possible to use one specifically molded for the application, or by the use of several layers of the material. One typical way in which excessive loading for a single section is handled is to cement two layers of the corrugated panel together with the corrugations crossed (Fig. 13-4). This results in a very stiff section capable of substantially greater weight bearing than a single sheet and it will meet the necessary requirements. The double sheet material also provides significant thermal insulation because of the trapped air space between the sheets.

The roof section was designed to meet the static load requirements. However, it is necessary to consider transient loads such as people walking on the roof and fluctuating wind loads. The localized loads represented by people walking on the roof can be solved as indicated in Table 13-2 by assuming concentrated loads at various locations and by doing a short time solution to the bending problem and the

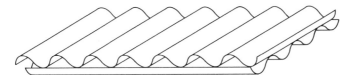

Fig. 13-4. Crossed corrugated sheets for high flexural stiffness.

extreme fiber stress condition. The local bearing loads and the localized shear should also be examined since they may cause possible local damage to the structure. Stresses from varying winds are general alternating stress loads and occur over wide areas of the structure. When the wind changes direction, the stress frequently changes direction, and the tendency is for the roof to lift away from the structure. The main point of stress caused by the wind is at the anchorage points of the roof to the rest of the structure. They should be designed to take lifting forces as well as bearing forces. Proper anchorage of the support structure to the ground is also essential. Local fire and building codes impose additional restrictions.

A large area of plastics, such as described here, has a substantial change in dimension with temperature. Surprisingly, very few of the traditional building materials, including wood, have significant expansion under normal temperature shifts. The fiber glass materials generally are not a problem since they have low thermal coefficients of expansion and the corrugated shape tends to flex and accommodate the changes caused by heating and cooling. In the case of materials such as vinyl siding, the expansion factor becomes significant and is an important consideration in the fastening system.

The effects of the environment on the performance of the material must be considered. Using the initial physical properties of the materials, the structure is sound. Exposure to weather, which includes water and sunlight, has a significant effect on the physical properties of the materials and this must be taken into account in the design. Table 13-2 gives some data on the change in physical properties of a typical polyester glass sheet molding compound (SMC) on exposure to the elements. It can be seen that there is a 50% or more drop in the physical properties in 5 years. This can be due to surface damage and to changes in the bulk of the material. In general, this type of loss of physical properties levels off to a low rate of deterioration in suitable materials so that any potential failure can be anticipated. This loss of properties can be compensated for by increasing the strength requirements by a suitable factor of safety, probably about three in this case, and by using a protective coating on the sheet material to minimize the effects of weathering. The preferred type of coating would be a fluorocarbon material which has the best resistance to sunlight and other weathering factors of all of the plastics. If this type of surfacing is used, the material will retain its surface integrity for 20 years.

The example of the roof structure represents the simplest type of problem in static loading in that the loads are clearly long term and well defined. Creep effects can be easily predicted and the structure can be designed with a sufficiently large safety factor to avoid the probability of failure. A seating application is a more complicated static load problem because of the loading situation. The self load on a chair seat is a small fraction of the normal load and can be neglected in the design. The loads are applied for relatively short periods of time of the order of 1 to 5 hours, and the economics of the application requires that the part be carefully designed with a small safety factor. A different design approach is used in this case. Instead of assuming an apparent modulus of elasticity using a constant creep situation covering the life of the chair, it is better to determine the actual creep deflection over a typical stress cycle, the creep recovery over a non-use cycle, and so on until the creep is determined after a series of what might be considered typical hard usage cycles for the chair. The accumulated creep after a period of 2 weeks can be assumed to represent the base line for an apparent modulus of elasticity to determine the design life of the chair.

With this basic approach in mind, let us do a design on a typical molded chair seat. The load will be assumed to be a 250 pound person and the typical load cycle is shown in Fig. 13-5, which includes loading times of 4 to 6 hours two or three times in 24 hours and a relaxation period of 1 to 2 hours during the day and of 10 hours during the night. The curve of loading is a random collection of these cycles over a 2-week period. The first step in the design is to select a section for

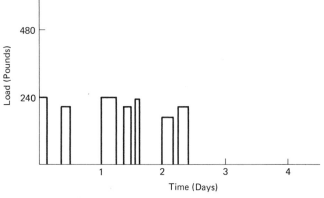

Fig. 13-5. Load time diagram for chair.

the chair seat which will have the required strength to prevent breakage with the stress calculated from the extreme fiber formula. The next step is to see that the seat does not deflect more than a given amount to be able to continue to function as a seat. An arbitrary deflection of 2 inches in the length of a seat 16 inches long will be assumed although such values are usually arrived at by consumer comfort testing. It might be noted in passing that in some chair designs where the creep did not result in failure of the chair, the fact that the seat was too resilient and gave a feeling of insecurity led to poor consumer acceptance. In many cases the "feel" of a product is important to its success and the feeling of solidity is important in furniture applications.

Within the limits set above the design can vary widely. The seat can be attached to the rest of the chair frame by leg supports at the four corners, or it can be cantilevered from the back with a floor pad support, or, in another version, from the front (Fig. 13-6). The seat construction can range from a formed sheet in two or three dimensions to one with rolled edges for reinforcement. It can have structural ribs molded in or it can be a sandwich panel construction made up of two molded parts cemented together. If can also be a structural foam molding, which is a sandwich panel construction. In each of the configurations there are tradeoffs of stiffness and strength which

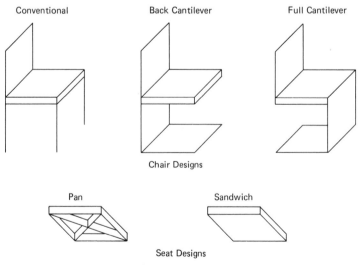

Fig. 13-6. Typical seating design.

may make one more effective than the others in meeting the seating requirements. In this case the designer has freedom of choice of both form and dimension as well as in the the selection of the materials. Given this freedom, it would be desirable to examine several of the alternatives to see which would provide the best seating at the lowest cost. Obviously, there is no point in doing all of the possibilities so a selection should be made on the basis of anticipated use as well as style requirements.

The three that will be analyzed here are shown in detail in Fig. 13-7, and a typical product is shown in Fig. 13-8. They are the single curve sheet cantilever mounted from the back, the molded pan supported on four legs, and the structural foam molding which is front supported. In order to simplify the analytical exercise, a particular material was selected for each. The single curved sheet is made of polyester fiber glass molded to the shape. The corner supported pan is molded from ABS plastics. The structural foam unit is molded from polypropylene with glass fiber filler. From inspection of the three designs it is apparent that the main stress of the loading will be at the

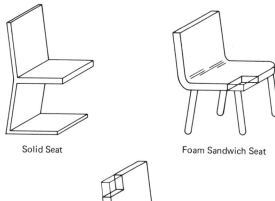

Solid Seat   Foam Sandwich Seat

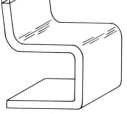

Foam Sandwich
Total Construction

Fig. 13-7. Figure for problem.

Fig. 13-8. These sturdy, nicely styled chairs are molded of reinforced plastics. (*Courtesy Herman Miller, Inc., Zeeland, Mich.*)

support point for the seat. This will be assumed to be sufficiently strengthened to prevent failure, either by excessive stress or bending at the support point. The analysis will be concerned with the fact that the seat itself will not break as a result of the load and will not sag excessively after continued use. For this example the impulse load caused by dropping into a chair will be ignored.

The deflection analysis is shown in Table 13-3. In each case the section is designed to keep the deflection to less than 2 inches in 16 inches for a design life of 5 years and the extreme fiber stress is kept to a value less than the yield strength of the material. The first step in the analysis is to determine the necessary section to resist the bending load using the short-term tensile and compressive strength and modulus values. The extreme fiber stress is calculated for these sections to determine that the part will not break when deflected. A time dependent modulus is then calculated using the extreme fiber stress level for each of the materials at the initial stress value level and using the loading-time curve developed above. If the deflection at the desired life is excessive, the section is increased in size and the deflection recalculated. By iteration the second can be made such that the creep and load deflection is equal to the maximum allowed at the design life of the chair. As indicated in Chapter 6, this calculation can be programmed for a computer solution.

The selection from the possible designs is made on a cost effectiveness basis. The least costly construction would be the best unless there is an inherently more useful construction for aesthetic or other reasons. In most design cases this will be an aesthetic consideration or, in this case, it may be a user consideration such as initial stiffness which leads to a better feeling of security. This may lead to the selection of something other than the lowest cost design. This value can be added to the consideration by examining the EI product for each of the constructions at the initial short-term loading. The higher the EI product, the less the seat will flex under load, and the more secure it will feel. With the materials and constructions examined, the initially stiffest construction may have a higher creep level and require a heavier and more costly construction to meet the creep criteria. This may be justified by the better feel of the seat. As always, the designer must be concerned with the utility of the product because this is his justification for designing it in the first place. The most important part of the design is that it satisfy the need for which it is made. The technical qualities must be such that they result in good acceptance. A technical success that leaves the user unsatisfied is not a product design success.

As with any design, the environmental and end-use requirements must be considered. The chair will be exposed to cleaning agents, children and dogs climbing on it, possible abuse in storage and shipment, etc. Such conditions should be considered as part of the material selection and design procedure. Of the three materials suggested, polypropylene has the best chemical resistance, ABS has the best abrasion resistance, and the polyester normally would have medium properties in both areas. All three are tough materials that would take rough handling, and the ABS would be best in terms of exposure to sunlight. In this particular design there would generally be no reason to choose any one of the materials over the other unless it is anticipated that the chair will have substantial exposure to conditions other than those typical of the home environment.

In the other design procedures we would tend to follow a minimum test sequence for acceptance testing in use to see that the design is functional, and set a reasonable quality testing procedure to insure that the processing is under control. While the use of plastics in a chair represents a situation where there is substantial personal injury risk it is one that is anticipatible and the normal design procedure would anticipate and eliminate the possibility of premature failure.

Deterioration of the chair with age should be examined to see if environmental exposure would lead to a shortened life. The indoor environment where a chair is normally used does not produce severe damage. Indoor sunlight is much less severe than outdoor exposure and room temperatures do not vary excessively. The only source of possible damage is in the use of cleaning and debugging agents that may attack the plastics. This can be controlled by following proper instructions. If difficulty is expected, a chemical resistant material is indicated. Abuse by dropping and impact should also be considered. This may cause surface or structural damage (see chapter 7). Public seating, as shown in Fig. 13-9, is subject to much abuse.

As examples we have examined two types of common statically loaded structural units. In each case they represent long-life expectancy units which will carry substantial loads. Since the failure of the part would involve substantial risk of personal injury, the

Fig. 13-9. Fiberglas-reinforced plastic school bus seats must be very rugged and easily cleaned. (*Courtesy Blue Bird Body Co.*)

designs must be done with caution. In one case, strong considerations are present and in the other traditional performance requirements must be considered. The loadings and the basis for making the design judgments are different. These two examples do not exhaust the possible combinations of conditions the designer will face, but they indicate what might be expected. The approach to the problem is to make the best analysis of the product requirements, including what at first appear to be intangible requirements, and then to determine what are the important elements in the design. Using these as the guide, several types of structural possibilities are examined with different materials to see if they meet the performance requirements of the application. The loads, the duration of the loads, the environment, and the intangible use factors will favor one design over the other, particularly if the economics are made the final basis for choice. The final design is made and tested for performance and sent to production with suitable quality control tests indicated. The process of design for static loads involves a great deal more than the mechanical operation of the stress strain data to determine the performance of a section. The results obtained from the stress analysis are used to determine the functionality of the product and then, combined with the other factors involved, to decide on a suitable design.

Using the approach suggested, designers can be guided in the design of static structure for performance in any environment from space, to aircraft, to land applications to subsea use. Defining the requirements and using the data available or generated for the application, the end result can be made predictable to a sufficient extent that successful products result with minimum cost.

Table 13-1 and Fig. 13-1A. Roof Problem Calculations.

The problem is to decide on which standard corrugated roof material made of reinforced plastics (glass-reinforced polyester) is to be used in a roof design with specified snow, wind, and live loads. The spacing of the supports is also analyzed. the problem is reduced to the bending of a section of continuous beam 1 ft wide with the supports placed at a specified point. The sag of the roof is limited to 4 in. for a design life of 1000 hours which is assumed to exceed the snow season for the climate considered.

The snow load is equal to 50 lb/ft^2.

Table 13-1. (Continued)

The wind load is ± 5 lb/ft^2.
The initial support distance is assumed to be 48 in.
The live load is taken to be 250 lb at the center of the span.
The standard corrugated sections are taken as intersecting semicircles as shown below and the moment of inertia of the sections is the same as a group of hollow cylinders to make the same equivalent developed length with the appropriate wall thickness. Two different spacing factor sections are examined 2 in./curve and 3 in./curve.

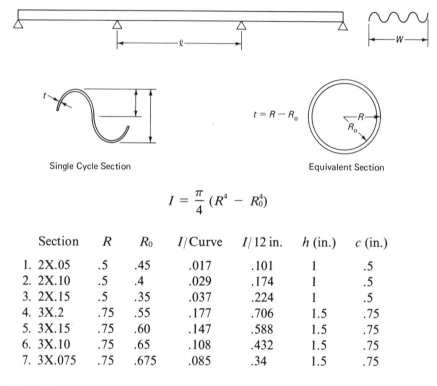

Single Cycle Section   Equivalent Section

$$I = \frac{\pi}{4}(R^4 - R_0^4)$$

Section	R	R_0	I/Curve	I/12 in.	h (in.)	c (in.)
1. 2X.05	.5	.45	.017	.101	1	.5
2. 2X.10	.5	.4	.029	.174	1	.5
3. 2X.15	.5	.35	.037	.224	1	.5
4. 3X.2	.75	.55	.177	.706	1.5	.75
5. 3X.15	.75	.60	.147	.588	1.5	.75
6. 3X.10	.75	.65	.108	.432	1.5	.75
7. 3X.075	.75	.675	.085	.34	1.5	.75

From Roark

$$M_{max} = \frac{W_1 l}{12} \qquad \delta_{max} = -\frac{1}{384}\frac{Wl^3}{EI}$$

for uniform load

$$M_{max} = \frac{W_3 l}{8} \qquad \delta_{max} = -\frac{1}{192}\frac{Wl^3}{EI}$$

Table 13-1. (Continued)

where

W_1 = snow load = 50 × 4 ft = 200 lb
W_2 = wind load = ±5 × 4 ft = ±20 lb
W_3 = live load = 250 lb

For max load at initial modulus $=E_1$ = 2,000,000 psi

$$\delta_{max} = \frac{1}{384} \frac{W_1 l^3}{EI} + \frac{1}{384} \frac{W_2 l^3}{EI} + \frac{1}{192} \frac{W_3 l^3}{EI}$$

$$\delta_{max} = \frac{1}{384 EI} (W_1 l^3 + W_2 l^3 + 2W_3 l^3)$$

$$= \frac{l^3}{384 EI} (W_1 + W_2 + 2W_3)$$

$$= \frac{\overline{48}^3}{384 \cdot 2000000 I} (50 \times 4 + 5 \cdot 4 + 250)$$

$$= \frac{\overline{48}^3 \cdot 470}{384 \cdot 2 \times 10^6 I} = 4. \text{ in.}$$

$$I = \frac{\overline{48}^3 \cdot 470}{384 \cdot 2 \cdot 4 \cdot 10^6} = .0170$$

An inspection of the table indicates that section 1 will be adequate for the elastic solution to the problem.

The maximum stress is calculated from the relationship

$$S_{max} = \frac{Mc}{I}$$

where

M = the maximum bending moment

$$M_{max} = \frac{W_1 l}{12} + \frac{W_2 l}{12} + \frac{W_3 l}{8}$$

$$M_{max} = \frac{200 \cdot 48}{12} + \frac{20 \cdot 48}{12} + \frac{250 \cdot 48}{8}$$

$$= 2380 \text{ in./lb.}$$

$$S_{max} = \frac{2380 \cdot .5}{.017} = 70,000 \text{ psi}$$

Table 13-1. (*Continued*)

which exceeds the max stress. Using $I = .101$ for section 1

$$S_{max} = \frac{2380 \cdot .5}{.101} = 11782$$

which also exceeds max stress. Using $I = .174$ for section 2

$$S_{max} = \frac{2380 \cdot .5}{.174} = 6839 \text{ psi allowable}$$

Only the snow load is a long-term load which will cause creep. The stress for this load is calculated and the appropriate apparent modulus is determined from the curve to check the required moment of inertia to check the selection of the appropriate section.

$$S_{max}(\text{snow load}) = \frac{200 \cdot \overline{48}^3 \cdot .5}{.174} = 2298.9 \text{ psi}$$

apparent modulus $= 1,070,000 \text{ psi} = E_A$

$$\delta_{max} = 4 = \frac{W_1 l^3}{384 E_A I} + \frac{W_2 l^3}{384 E_1 I} + \frac{W_3 l^3}{192 E_1 I}$$

$$4 = \frac{200 \cdot \overline{48}^3}{384 \cdot 1.07 \cdot 10^6 I} + \frac{20 \cdot \overline{48}^3}{384 \cdot 2 \cdot 10^6 I}$$

$$+ \frac{250 \cdot \overline{48}^3}{192 \cdot 2 \cdot 10^6 I}$$

$$I = (13,458 + 720 + 18,000) \times 10^{-6} = .032$$

Therefore section 2 is okay

$$\delta_{actual} = \frac{200 \cdot \overline{48}^3}{384 \cdot 1.07 \cdot 10^6 \cdot .174} + \frac{20 \cdot \overline{48}^3}{384 \cdot 2 \cdot 10^6 \cdot .174}$$

$$+ \frac{250 \cdot \overline{48}^3}{192 \cdot 2 \cdot 10^6 \cdot .174}$$

$$= .7397 \text{ in.}$$

Table 13-2. 15-Year Performance Experience with SMC.

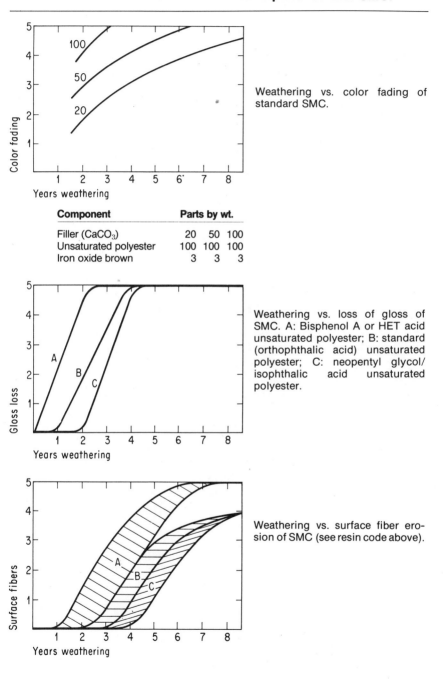

Weathering vs. color fading of standard SMC.

Component	Parts by wt.		
Filler (CaCO₃)	20	50	100
Unsaturated polyester	100	100	100
Iron oxide brown	3	3	3

Weathering vs. loss of gloss of SMC. A: Bisphenol A or HET acid unsaturated polyester; B: standard (orthophthalic acid) unsaturated polyester; C: neopentyl glycol/isophthalic acid unsaturated polyester.

Weathering vs. surface fiber erosion of SMC (see resin code above).

Table 13-2. (Continued)

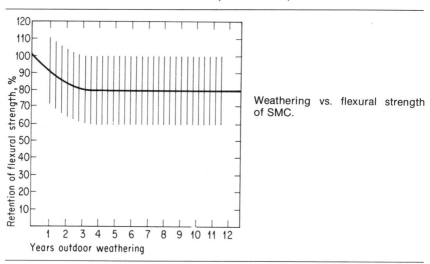

Weathering vs. flexural strength of SMC.

(Courtesy Modern Plastics, May 1975)

Table 13-3. Chair Problem

This problem is based on the use of a chair seat which is supported as a cantilever beam at the rear. The person load is assumed to be central on the seat and the controlling requirement is that the sag of the seat not exceed 2 in. for the life of the chair which is assumed to be 4 years. The components of the deflection are the stress induced deflection from the load plus the creep caused by the fact that the seat does not completely recover between applications of the stress. The assumed cycle of use is 4 hr on 1 hr off 4 hr on 15 hr off and this cycle is repeated daily. Two different seat constructions are examined. One is a glass-filled polypropylene molded section and the other is a foam core sandwich panel. In each case the solutions are not intended to be exact since many of the data are not available but as an indication of the effects of the main variables on the design and as a starting point for a design that will probably work.

The problem is shown in Fig. 13-1. The load is 250 lb and the seat size is 16 × 16 in. The loading cycle is shown in Example 13-1.

a = 8 in.
b = 16 in.
l = 16 in.
W = 250 lb
h = thickness
$c = \dfrac{h}{2}$

Table 13-3. (Continued)

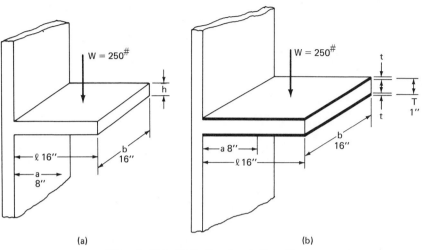

(a)   (b)

Example 13-1. Illustration for chair problem.

The material is glass-filled polypropylene

initial elastic modulus = 900,000 psi

and the apparent modulus is shown in Example 13-2. The elastic solution is:

$$\delta_{max} = \frac{1}{6} \frac{W}{EI} (3a^2l + a^3) \qquad \text{(from Roark)}$$

$$2 \text{ in.} = \frac{1 \cdot 250}{6 \cdot 900000 \cdot I} (3 \cdot 8^2 \cdot 16 - 8^3)$$

$$2 \text{ in.} = \frac{250 \cdot 3}{6 \cdot 900000 \cdot 4 \cdot h^3} (3072 - 512)$$

$$I = \frac{bh^3}{12} = \frac{4h^3}{3}$$

$$h = \sqrt[3]{\frac{.0884}{2}} = .354 \text{ in.}$$

$$I = \frac{4h^3}{3} = \frac{4 \cdot .385^3}{3} = .076$$

$$M_{max} = Wl = 250 \cdot 8 = 2000 \text{ in. lb}$$

$$S_{max} = \frac{Mc}{I} = \frac{2000 \cdot \frac{.385}{2}}{.076} = 5065.8 \text{ psi}$$

Table 13-3. (Continued)

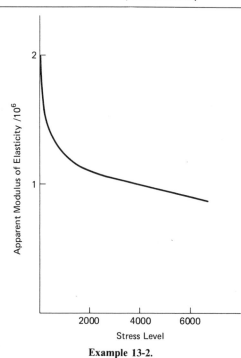

Example 13-2.

From the curve the unrecoverable loss in deflection results from the fact that the modulus recovers only to 850,000 psi for a stress level of 5056 psi.

For the analysis of this problem this is the apparent modulus which is used to determine the long-term deflection of the seat. When this is added

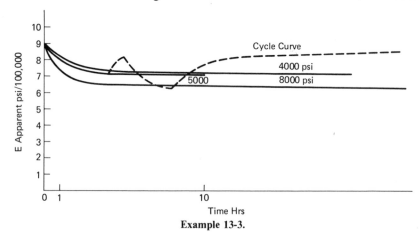

Example 13-3.

Table 13-3. (Continued)

to the effect of the initial load applied over the initial cycle the deformation of the seat is obtained.

The apparent modulus for the 4-hr load is 710,000 psi

$$\delta_{max_{4hr}} = \frac{1}{6} \frac{W}{EI} (3a^2l - a^3) = 2 \text{ in.}$$

$$= \frac{1}{6} \frac{.250}{710000 \cdot I} (2560) = 2 \text{ in.}$$

$$I = .075$$

The unrecoverable deflection =

$$\frac{E \cdot E_L I}{E} \cdot 2 \text{ in.} = \frac{900000 - 850000}{900000} \cdot 2 \text{ in.} = .11 \text{ in.}$$

$$\delta_{max_{4yr}} = 2 - .11 = 1.89 \text{ in.} = \frac{250 \cdot 2560}{6 \cdot 710000I}$$

$$I = .0795 = \frac{4h^3}{3} \qquad h = 391 \text{ in.}$$

$$S_{max} = \frac{2000 \cdot \dfrac{.391}{2}}{.0795} = 4914 \text{ psi}$$

Since this value of the maximum stress is close to the value assumed to obtain the apparent modulus the answer is correct within the accuracy of the data. If it were not, the new apparent modulus is obtained by interpolation between the curves and the process repeated until the assumed stress level and the corresponding apparent modulus are close to the assumed values. This calculation can be set up for computer analysis and the appropriate flow chart and program are shown below. (These data will be meaningless to persons who are unfamiliar with computer technology.)

In the second analysis of this problem a sandwich panel construction with polyester glass skins and a foam core is used. For purposes of this problem the core is assumed to be strong enough to prevent skin buckling but does not contribute to the stiffness of the beam. The chair dimensions are the same as for the other solution. The analysis will be limited to a determination of the skin thickness for a sandwich panel spacing of 1 in. from the center of one skin to the center of the other. The beams are shown in Example 13-1 (page 249).

Table 13-3. *(Continued)*

$a = 8$ in.
$b = 16$ in.
$l = 16$ in.
$T = 1$ in. center to center of skins with thickness $= t$
$A = $ section area $= 16 \cdot t$

from Roark

$$I = I_{\text{skins}} + A \frac{T}{2}$$

$$I = 2 \cdot \frac{1}{12}bt^3 + A\left(\frac{T}{2}\right)^2$$

$$I = \frac{2 \cdot 16 \cdot t^3}{12} + 16t\left(\frac{1}{2}\right)$$

$$I = 2\frac{2}{3}t^3 + 4t$$

$$\delta_{\text{max}} = -\frac{1}{6}\frac{W}{EI}(3a^2l - a^3)$$

$$E_{\text{initial}} = 2.2 \times 10^6$$

$$\delta_{\text{max}} = 2 = \frac{250}{6 \cdot 2.2 \times 10^6 I}(2560)$$

$$I = .024242 = \frac{8t^3}{3} + 4t$$

Since $t \ll 1$ for first estimate neglect $8t^3/3$

$$t = .00606 \quad \frac{8t^3}{3} \ll 10^{-8}$$

thus it can be neglected

$$S_{\text{max}} = \frac{MC}{I} - M_{\text{max}} = 200 \text{ in. lb}$$

$$S_{\text{max}} = \frac{2000 \cdot \frac{1}{2}}{.024242} = 41250 \text{ psi}$$

Table 13-3. (*Continued*)

This exceeds max stress for material

max allowable stress on material = 10000 psi

$$10000 = \frac{Mc}{I} = \frac{2000 \cdot \frac{1}{2}}{I}$$

$$I = \frac{2000 \cdot \frac{1}{2}}{10000} = .1 = 4t$$

$$t = .025 \text{ in.}$$

$$\delta_{max} = \frac{1}{6} \frac{W}{EI} (3a^2 l - a^3)$$

$$= \frac{250 \cdot 2560}{6 \cdot 2.2 \cdot 10^6 \cdot .1} = .4848 \text{ in. elastic solution}$$

Find $E_{apparent}$ δ_{max} = 2 in.

$$2 = \frac{250 \cdot 2560}{6 \cdot E_a \cdot .1} \qquad E_a = \frac{250 \cdot 2560}{2 \cdot 6 \cdot .1} = 533333 \text{ psi}$$

the solution is adequate.

The apparent modulus for the 4-hr load is 1.4×10^6 psi and the deflection after a cycle is

$$\delta_{act} = \frac{1}{6} \frac{250 \cdot 2560}{1.4 \times 10^6 \cdot .1} = .762 \text{ in.}$$

The solution of this problem is not creep dependent but stress dependent. If the values that resulted from the initial elastic solution indicated stress levels less than the strength of the material a stepwise solution such as was done for the glass-filled polypropylene is used and the computer program can be used to calculate the skin thickness.

Chapter 14

Design of Dynamically Loaded Plastics Parts and Evaluation Procedures

Parts used in machine applications to transmit motion are subjected to varying loads. This is the major use of dynamic loading although dynamic loading occurs in structures which are exposed to severe vibration forces and in structures where the flow of a fluid causes excitation of surface motion by cavitation and flow effects in the fluid. In each case the type of load that is encountered is usually of short time duration with time constants under one minute. The loads are cyclical in nature with each loading cycle essentially the same. In most cases the parts subjected to the cyclical stress either are in motion or are being excited into motion. Consequently, inertial forces are affecting the loading of the part.

The discussions in Chapter 6 and in the preceding material have shown that the stress-strain characteristics of plastics materials are very highly dependent on the rate of application of the load. The responses at high loading rates tend to be more elastic than viscoelastic and the time rate of loading and recovery is such that little or no creep occurs. In terms of the models, the response is that of a spring and dashpot combination where the time constant of the dashpot is of the order of seconds or less. The stress-strain curves are nonlinear in most cases, but the strain recovery after a loading cycle is essentially complete. Figures 14-1 and 14-2 are typical stress-strain curves in loading and unloading for a plastic material. The curves are shown for one loading rate and several peak stress levels. The dotted line shows how the curve would be changed for a different loading rate.

There are two types of designs involving dynamic loading. One is the rigid body problem which involves power transfer by the action of the member and the other is the use of a plastics member as an energy absorption member for damping purposes. The basic difference be-

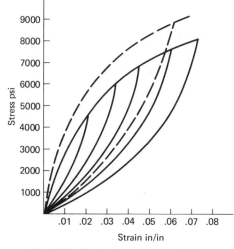

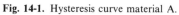

Fig. 14-1. Hysteresis curve material A.

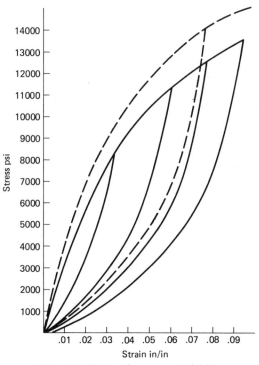

Fig. 14-2. Hysteresis curve material B.

tween the two types of applications is in the manner in which the hysteresis energy is treated in the problem. In the energy transfer situation, the effect of the energy conversion to heat should be minimized to prevent part failure and to improve the efficiency of energy transfer. In the case of the damping applications, the energy absorption is enhanced to improve the damping effects and the transmitted energy is minimized. Both types of problems are concerned with the generation of heat by the hysteresis effects. In all of the designs the removal of the frictional heat is a major part of the design problem in order to prevent failure by excessive heat build up.

The elements of the design procedure for a dynamically loaded part are as follows:

1. Determine the static loads present on the part in addition to the dynamic loads. These must have their effects added to the imposed loads on the part.
2. Determine the dynamic loads applied to the part in magnitude and duration. These loads will be compressive, tensile, bearing, bending, and shear loads and, in many cases, combinations of the loads will be present. The time constant of the loadings must be determined. If possible a load time curve should be constructed for the manner of loading and if the load is cyclical (such as it would be in a continuous machine use), the frequency of the loading should be determined.
3. In most machine applications the part under consideration is in motion, usually in a complicated manner. This motion indicates that the part is being accelerated periodically and that inertial loads are applied to the part by these accelerations. The magnitude of the stresses generated by these inertial loads must be calculated and added to the total stress applied to the part. Frequently there are cases where the inertial loads are the primary load on the part, and neglect of them can lead to failure of the part.
4. Since one of the important parts of the design of a dynamically loaded part is concerned with the heat generated by the hysteresis effect, all of the relevant factors concerned with heat transfer from the part should be determined. This includes the thermal properties of the material such as heat capacity, thermal conductivity, heat distortion temperature; the surroundings such as the adjoining structures to which heat can be transferred by conduc-

tion; and the surrounding media such as air or liquid to which heat can be transferred by convection, and possibly heat transfer by radiation (although at the operating temperatures for most plastics, radiant heat transfer is relatively unimportant); and finally, the structural variations possible to improve the heat transfer situation.

5. Determine the level of risk involved in the part. Obviously a helicopter is a more critical application than the driving link in a toy automobile but, between these extremes there is a specific hazard level associated with failure. The mode of failure and the possible danger associated with the failure must be evaluated to determine the criteria for the design performance.

6. Determine the effect of the environment on the performance of the part. Since the heat dissipation is critical in dynamic applications, it is essential that the thermal environment be determined as accurately as possible, including the possibility of a temperature extreme that would result from abnormal use. The effect of such thermal excursions on the part performance should be carefully considered, especially in critical applications. The effect of other environmental factors on the degradation of physical properties should also be considered to determine the design life.

7. The result of fatigue effects on the part should be determined. In general this is the life limiting effect in parts which are designed to avoid the effects of overheating and of environmental deterioration.

8. Devise tests to evaluate life performance and the effects of processing on part quality. Life test evaluations for everything but fatigue are comparatively simple for the dynamic loading since the time dependent creep problem is not present. Fatigue effects can only be determined as a statistical effect and the S-N curves generated used as a guide to safe stress levels. There is no reasonable way that this can be accelerated and if fatigue effects are critical, values should be generated for the materials to be used before design is undertaken.

In order to illustrate the principles of design, a problem will be used which incorporates the design elements involved. In Fig. 14-3 a cam and link arrangement is shown which was to be used in the operation of a hedge cutter. The link transmitted the power from the

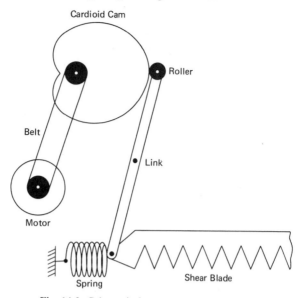

Fig. 14-3. Schematic for the hedge shear problem.

rotating cam to the scissor section of the hedge cutter and acted against the spring return on the cutter blades. Both the cam and the link are dynamically loaded and between them they represent most of the types of loading present. The cam was driven by the motor through a belt which adds the tensile loading and bending modes of intermittent loading. By including this in the overall analysis the problem encompasses most modes of dynamic loading.

The parts are shown in Figs. 14-4, 14-5 and 14-6. The loading from dynamic stresses is shown in Fig. 14-7. The curves in Figs. 14-1 and 14-2 give the stress-strain relationships for the materials involved. The type of materials used for these parts would be glass-filled plastics such as styrene or polyester for the link, thermoplastic polyester or

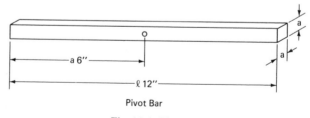

Fig. 14-4. Pivot bar.

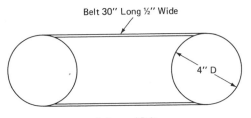

Pulley and Belt
Fig. 14-5. Pulley and belt.

nylon for the cam, and a nylon-reinforced rubber belt for the drive member. The curves are arbitrary ones for the presumed materials because the required data are not readily available for real materials.

The analysis will start with the link. This is a simple lever which is a beam. The loads are the driving load from the cam, the static load from the spring in the fully relaxed position of the lever, the dynamic load from the spring as the link operates, and the inertial load produced by the pivoting of the link. Using the Mc/I relationship, the

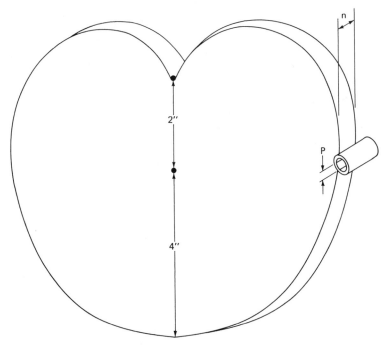

Fig. 14-6. Hedge cutter illustration cardioid cam.

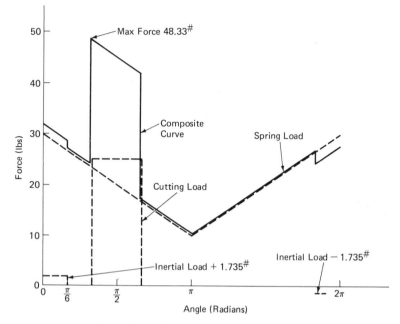

Fig. 14-7. Load time diagram for hedge cutter.

extreme fiber stress can be calculated for a specific beam geometry. Using a stress level permitted by the stress strain curve, the necessary beam cross section can be calculated. While a number of different beam cross sections can be considered, an *I* beam shape is assumed because it would represent an efficient section for this application, particularly in view of the fact that the pivot will tend to weaken the section. In the square beam shape the pivot will have a minimum effect on the reduction of the effective cross section.

Given the specific stress level chosen, several things must be calculated. The bearing load at the pivot point is calculated to see if it is not excessive for the material involved. If it is, the pivot area might be built up to reduce the stress level or the diameter of the pivot increased to reduce the stress level. The effect of the static spring load must be considered. This load on the cross section chosen should be examined for long-range creep effects which will be superimposed on the cyclical stress caused by the actuation of the lever. In all probability this will be minor since the spring force level will be small compared with the operating stresses.

The next step in the analysis is to determine the inertia loads that act on the link and also the part that it will drive. The link will be a relatively low inertia element but the blade that it drives will be a substantial inertial load. The value of these loads are calculated and added to the direct spring load on the link. This will probably necessitate increasing the section of the beam to take the inertial loads.

The load time diagram for the link is plotted in Fig. 14-7 at the point of maximum stress which would be at the outer fibers of the beam. Using the stress levels and loading rates that can be determined from the load time diagram, it is possible to find the desired stress-strain curve which represents the loading rate and stress level for the part at the expected operating temperature. Since the peak stress level is the critical one, that curve should be used to find the hysteresis level. If the curves for the stress-strain relationship can be written in explicit form, the area between the stressing and stress relieving curves can be integrated to obtain the area between the curves over the stress range used. If not, the area can be measured with a planimeter or by counting squares. The area will represent to the scales of the graph the energy absorbed per stressing cycle. The product of this value and the number of cycles of loading per unit time will give the rate of heat generation. This is the heat that must be dissipated in order to prevent overheating of the part with a consequent failure.

The heat transfer rate from the link must be determined. In order to do this a heat transfer coefficient must be estimated, the heat transfer area (usually the surface area of the part) calculated, and the average temperature difference between the part and the surroundings determined. The heat transfer equation is

$$q = UA(T_{part} - T_{surroundings})$$

where

U = heat transfer coefficient
A = heat transfer area
T = temperature

The value for U can be determined from a table of heat transfer coefficients as listed in a heat transfer text such as MacAdams or Jacob or from the Chemical Engineers Handbook. These will be given for a similarly shaped body with natural convection and with forced convection at several air velocities. Using the surface area of the part and the U values from the reference, calculate the heat transfer from the part. The calculations are done for several estimated part temperatures, for the still air, and the forced convection condition at several air velocities. From the estimated heat transfer rates find the set of conditions where the heat transfer is equal to or greater than the internal heat generation. In each case when a different part temperature is assumed, the stress-strain curves and hysteresis loop for this temperature must be used to determine the internally generated heat. It may be that at the stress levels initially assumed, there is no solution. The internally generated heat may increase faster than the rate at which the heat can be removed by the increased temperature driving force even with circulating air at reasonable velocities. If this is the case, the cross section is redesigned to reduce the stress level and the design procedure repeated until there is a temperature that the part can reach where the heat generated by the internal hysteresis is removed by the convection heat transfer to the surroundings.

It should be noted that this is a conservative solution to the problem for this type of part. Only the extreme fibers in the beam and only at one point in the link does it reach the value used to estimate the hysteresis energy generation. The inner parts of the link are stressed to a much lower degree and, consequently, generate much lower levels of frictional heat. As a result, the solution obtained is a safe one provided that the surrounding temperature is not higher than assumed. This condition would upset the heat transfer estimate. The iterative solution to this problem can be programmed for computer solution. The curves for stress-strain at elevated temperature can be generated from the room temperature curves using the *WLF* relationship. The only cross check that should be made is that at the equilibrium part temperature, the creep level from the sustained spring load should not be excessive. It should be calculated for the final beam shape and the equilibrium part temperature.

The solution to the link problem may require that the air be circulated across the link. This can be done by putting air movers on the link, on the cam, on the blade assembly, or on all of the parts. The cam can be made with a propeller center section so that it will act like a fan. This cooling action would be helpful in making all of the parts perform under the dynamic loading.

The next part of the assembly to be examined is the cam. The pins connecting the link to the blade, the pin at the pivot, and the pin which rides on the cam are all assumed to be made of metal so that they will not be excessively stressed by operation of the device. The cam design is a cardioid cam which has two Archimedes spirals that join at the high and low points of the curves as shown in Fig. 14-6. This cam gives simple linear motion to the pin driving the link which is translated into similar motion on the blade. The only points of acceleration are at the turning points in the curves where the motion goes from one direction to the other. Along the rest of the curve the pin pressure on the cam is caused by the displacement of the spring. A curve of the forces acting on the cam, including spring force plus the inertial forces caused by the change in direction of the blade and link is shown in Fig. 14-7.

The stress applied to the cam by the drive pin is a compression force that moves along the surface of the cam groove, plus the friction reaction caused by the pressure and movement. The force will be slightly tilted from normal to the surface of the part. With the material selected, the coefficient of friction is small so that the normal force will be essentially the only force applied to the cam. Our analysis will be concerned with the stress level applied by the driving pin to the cam at the reversal points of the cam since these represent the points of maximum stress. With an assumed area of contact and the forces calculated from the spring forces and the calculated inertial forces, the compressive stress level can be calculated. Using the stress-strain relationship for the cam material, the hysteresis loop energy per cycle can be measured. By multiplying this by the number of cycles per unit of time the rate of frictional heat generation can be calculated.

We use the same analytical approach to the problem as we did before. The bearing area must be designed such that the maximum stress generated will be a reasonable fraction of the maximum working stress of the material. The stress-strain curve in compression used for the initial estimate of the heating is based on this value. The heat transfer from the point of maximum stress must be calculated, but in this case the conduction of heat from the point of high stress to the remainder of the part will be a major part of the heat transfer. This was not so in the case of the link. Jacob gives a solution for heat transfer from a localized heated point to the bulk of a large part. This solution combined with the solution for convection heat transfer from the surface of the part will give the rate at which heat can be dissipated from the part at several assumed operating temperatures for the maximum stress area. Using this in conjunction with the hysteresis loop for the assumed temperature, we can find a temperature and air circulation requirement at which the internal heat generated will be dissipated by heat transfer to the surroundings. As in the case of the link, if there is no solution at the assumed stress level because the frictional heat generated at the higher temperatures exceeds the increased heat transfer, the problem is reworked at a lower stress level and repeated until the stress level is such that the part will assume an equilibrium temperature that will not cause catastrophic failure.

This solution must also be examined for the effect of compression creep caused by the constant spring tension. It might be pointed out that while the solution obtained is a safe solution provided that the surrounding temperature is not different than that assumed, it is not a conservative solution. The reason is that a small section of the part is being heated by the hysteresis effects and it must be quickly rid of the heat. This depends on a more complicated heat transfer situation which can change and lead to the start of melting breakdown which will cause rapid failure. In cases like this a safer approach might be to use the cam immersed in a liquid heat transfer medium, but it is impractical in this design. In this situation the use of air circulation created by the part movement is the most effective approach. Another approach would be the use of a thermosetting material with higher heat resistance.

Table 14-1 (pp. 265–271) contains the equations and solutions for the cam section of the design.

The last unit in our design problem is the drive belt. If the cutter were electrically driven by a constant speed motor, the major function of the belt would be the uniform transmission of power to the cam and, subsequently, to the cutter head. In this design, the cutter is driven by a gasoline motor which has pulsations in power output particularly a small motor of the type used in hand held units. The amount of flywheel effect that can be built into such a unit is limited and the pulsations caused by the individual firings in the motor cylinders is transmitted to the output shaft.

In this design an additional function of the drive belt is to reduce the impulse loading transmitted from the motor. The belt selected is

Table 14-1. Hedge Cutter Problem.*

To illustrate cyclical loading of plastic parts and the design limits we will examine the parts of a hedge cutter the construction of which is shown in Fig. 14-3. The separate basic elements are shown in Figs. 14-4, 14-5 and 14-6. The cam is a cardioid cam with constant radial velocity and constant radial acceleration at the turn around points. The cutter section, which is assumed to be steel, is driven by the cam through the lever arm and the restoring force is supplied by a spring. The spring has a force equal to 10 lb times the displacement. The spring is preloaded to 10 lb force. The cutter has a mass of 5 pounds. The lever length is 12 in. and it is centrally pivoted with a roller contacting the cam and a pin connecting it to the cutter bar. The cutting load is assumed to be 25 lb and it acts on the forward stroke of the cutter. The cam rotates at 100 rpm. The first item to be calculated is the inertial load and for this only the cutter bar is considered as significant.

$$r = \frac{2\theta}{\pi} + 2\Big|_0^\pi \qquad r = \text{radius} \qquad \theta = \text{angle} \qquad (1)$$

curve of cam displacement per half revolution = 2 inches

$$\text{radial velocity} = \frac{2 \text{ in.}}{\dfrac{1}{200} \text{ min}} = 400 \text{ in.}/\text{min} = 6.67 \text{ in.}/\text{sec}$$

$$\frac{2\pi \cdot 100}{60} = w = 10.47 \text{ rad}/\text{sec}$$

this becomes 0 in $(\pi/6)$ rad at constant acceleration

$$\frac{(\pi/6) \text{ rad}}{10.47 \text{ rad}/\text{sec}} = .05 \text{ sec}$$

$$a = \text{acceleration} = \frac{\dfrac{6.67 \text{ in.}/\text{sec}}{12}}{.05 \text{ sec}} = 11.11 \text{ ft}/\text{sec}^2$$

$$F = Ma = \frac{5 \text{ lb}}{32.2} \cdot 11.11 = 1.735 \text{ lb inertial force}$$

Spring force = 10 to 30

Max F_{tot} = 1.735 + 30 = 31.735 lb

Min F_{tot} = −1.735 + 10 = 9.265 lb

*Portions of this presentation will be meaningless to persons who are unfamiliar with computer technology. The computer analysis was developed by Lloyd E. Levy.

Table 14-1. *(Continued)*

The load time diagram for the ends of the lever arm is shown in Fig. 14-7. The maximum force is at $\pi/3$ radians and is equal to 48.33 lb.

$$M_{max} = \frac{1}{4} WL = \frac{2 \times 48.33 \times 12}{4} = 289.98 \text{ in. lb}$$

material is material B shown in Fig. 14-2. The lever is a square section dimension a max operating temperature = 170° F. environment = 70° F. heat transfer coefficient = 10 btu/hr/ft^2/° F.

Table of Values

a	a^2 Area	$12a^2$ Vol.	Mc/I Max Stress	Energy/Cycle	Energy Rate Btu/hr	Heat Transfer Area	Heat Transferred 100° F Temp. Dif. Btu
.5	.25	3	12000	250	7712	.167	167
.6	.36	4.32	6944	35	1080	.2	200
.7	.49	5.88	4373	15	463	.23	230
.8	.64	7.68	2929	5	154	.267	267
.9	.81	9.72	2058	2	61.7	.3	300
1.0	1.0	12	1500	1	36.85	.333	333
1.1	1.21	14.52	1126	.6	18.5	.367	367

$$c = \frac{a}{2}$$

$$I = \frac{a^4}{12}$$

$$\text{Max stress} = \frac{289.98 \cdot \dfrac{a}{2}}{\dfrac{a^4}{12}} = \frac{289.98 \cdot 6}{a^3}$$

$$\text{Energy rate} = \frac{\text{energy}}{\text{cycle}} \times \frac{\text{area} \times l \times 400 \times 60}{12}$$

$$= \frac{\text{ft lb/hr}}{778} = \text{Btu/hr}$$

$$= E/c \cdot A \cdot 30.85 \text{ Btu/hr}$$

$$\text{Heat transfer area} = \frac{4al}{144} = \frac{48a}{144} = \frac{a}{3}$$

The foregoing tabulation shows the values of internal heat generated based on an integration of the hysteresis curves from Fig. 14-1 which is shown as

Table 14-1. *(Continued)*

a curve vs maximum stress in Fig. 14-2. The various sizes of the bar are shown to give certain heat transfer areas which use an assumed heat transfer coefficient of 10 and a temperature difference of 100° F between the part and the environment. At a cross section with $a = .7$ to $.8$ the curves are assumed to be taken at 170° F. Since we are assuming that the entire volume of the bar is stressed to the maximum stress level for the entire duration of the cycle on and off for the unloading cycle the assumed energy generation is substantially greater than the actual case. The part is then conservatively designed. To make a complete solution, curves of energy dissipation vs stress level are required for a range of material temperatures, exact heat transfer data and mechanisms are necessary, and the solution should be tested for several levels of ambient temperature. This would lead to a more refined solution but would only be justified in a case where the data are readily generated, and the test of the final design would be costly and difficult. The solution above is a safe overdesign condition provided that the ambient temperature does not materially depart from 70° F. Even at 100° F the table would indicate that the 0.8 in. a section would be acceptable.

The cam used in the hedge cutter is analyzed next. The cam is shown in Fig. 14-6. It is assumed that an element of the cam p long/n width of the cam wide is loaded in compression by the roller engaging the cam attached to the pivot bar. For purposes of estimating the hysteresis energy generation the stress is assumed to act to a depth of 1 inch. For the entire revolution of the cam the surface 1 in. layer is loaded in compression to levels ranging from 10 to 48.33 lb. This is equivalent to stress levels of $48.33/pn$ to $10/pn$. The contact area can be estimated based on engineering formulas given the physical constants of the materials, and, in this case it is taken as 0.01 in. The stress levels then become stress $= 4833/n$ to $1000/n$. For a conservative analysis it is assumed that the maximum stress level is reached at all points on the surface and this is the equivalent of having an area equal to n times the perimeter stressed once per revolution. The perimeter of the cam is equal to 6π. The data for material B shown on the curves of Fig. 14-2 will be used. The heat transfer coefficient $H_k = 10$ Btu/hr/° F and the temperature difference will be $170° F - 70° F = 100° F$.

n Cam Width	S 4833/n	Energy/Cycle	Energy Rate Btu/hr	Heat Transfer Area $6\pi n$	Heat Transferred Btu/hr
.5	9666	185	2241	.065	65
.6	8055	90	1090	.078	78
.7	6904	45	545	.092	92
.8	6041	30	363	.104	104
.9	5370	15	182	.118	118
1.0	4833	10	121	.130	130

Table 14-1. (Continued)

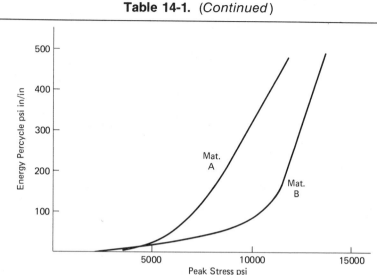

Fig. 14-8. Hysteresis energy per cycle vs. peak stress at 400 cycles/min. materials A and B.

$$\text{Energy rate Btu/hr} = \frac{\text{Energy/cycle} \times 6\pi \times 100 \times 60}{12 \times 778}$$

$$= 12.114n \times \text{Energy/cycle}$$

The values for the energy rate are arrived at by integrating the hysteresis loop curves which are shown as a curve in Fig. 14-8 as a function of the stress level.

It is apparent by inspection that the internal heat generated is equal to the heat transferred when the cam width is equal to between .9 and 1.0 in. The answer is conservative since it assumes that the entire surface of the cam is stressed to the maximum stress level when in fact only one point on the cam is exposed to the maximum stress. In addition the energy generation is based on maximum temperature curves when in fact most of the cam is at a lower temperature and the corresponding energy generation would be less. A particular stressed volume is used which may in fact be larger than the actual stressed volume and the stress distribution is quite complex, which may make the solution less conservative. The heat transfer situation is limited to heat lost from the outer periphery of the cam and it is based on the unlikely assumption that the entire volume of the cam is at the same temperature. Both of these under estimate the amount of heat removed. As with the analysis on the link this is a crude estimate of the actual situation and is useful as a first approximation on the conservative side. The use of more refined heat transfer relationships, data reflecting the internal energy generation at several temperatures, and a more accurate stress concentration analysis

Table 14-1. (Continued)

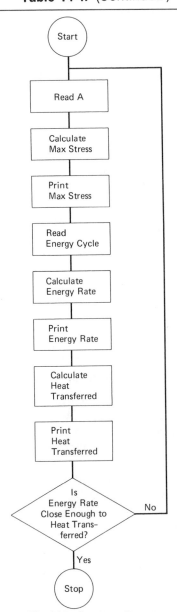

Fig. 14-9. Computer flow chart.

would give a much more accurate solution. The solution as given would be a useful starting point for testing and would be a safe design if more detailed analysis is not justified by the application.

Table 14-1. *(Continued)*

```
     ∇BAR[□]A
   ∇ BAR
[1]  LOOP:'SECTION SIZE'
[2]  A←□
[3]  'ENERGY/CYCLE AT STRESS OF ';1739.88÷A*3;' IS'
[4]  EC←□
[5]  'ENERGY RATE IS ';EC×30.85×A*2;' BTU/HR, HEAT TRANSFERRED IS ';333.3×A;' BTU/HR'
[6]  →LOOP
   ∇

     BAR
SECTION SIZE
□:
     .5
ENERGY/CYCLE AT STRESS OF 13919 IS
□:
     525
ENERGY RATE IS 4049.1 BTU/HR, HEAT TRANSFERRED IS 166.65 BTU/HR
SECTION SIZE
□:
     .6
ENERGY/CYCLE AT STRESS OF 8055 IS
□:
     53
ENERGY RATE IS 588.62 BTU/HR, HEAT TRANSFERRED IS 199.98 BTU/HR
SECTION SIZE
□:
     .7
ENERGY/CYCLE AT STRESS OF 5072.5 IS
□:
     25
ENERGY RATE IS 377.91 BTU/HR, HEAT TRANSFERRED IS 233.31 BTU/HR
SECTION SIZE
□:
     .8
ENERGY/CYCLE AT STRESS OF 3398.2 IS
□:
     7
ENERGY RATE IS 138.21 BTU/HR, HEAT TRANSFERRED IS 266.64 BTU/HR
SECTION SIZE
□:
     →
```

Fig. 14-10. Computer data.

It is apparent that given the energy generation data in tabular form the problem can be solved by computer. This would be a particularly desirable step if data at several temperature levels are available and the effect of this on the solutions taken into account. A flow chart and program for this follows (Figs. 14-9 and 14-10).

The torque on the drive belt is next analyzed. The ramp angle on the cam is equal to the rise of 2 inches divided by the peripheral distance which is 3π. The angle of rise equals arc tan $2/3\pi$. The maximum torque will be at the point where the force is equal to 48.33 pounds when $\theta = \pi/3$ at which point the radius is equal to 2.67 inches.

$$\text{Maximum torque} = \frac{2 \times 48.33 \times 2.67}{3\pi} = 27.83 \text{ in./lb}$$

We will assume a driving and driven pulley 4 in. in diameter so that the tension in the driving belt will be $27.83/2 = 13.7$ lb. The belt is assumed to be made of a material A (Fig. 14-1), and is $1/2$ in. wide. The heat transfer coefficient $H_k = 10$ Btu/sq ft/° F and the temperature difference is again 100° F. The belt thickness is q.

Table 14-1. (Continued)

q Belt Thickness	Stress Area	Stress Max. psi	Energy/Cycle	Energy Generation Rate Btu/hr	Heat Transfer Area	Heat Transfer Btu/hr
.002	.001	13700	475	91.6	1.004/144	6.97
.004	.002	6850	26	10	1.008/144	7
.006	.003	4567	15	8.7	1.012/144	7.03
.008	.004	3425	7	5.4	1.016/144	7.05
.010	.005	2740	5	4.8	1.02/144	7.08
.012	.006	2283	3	3.5	1.024/144	7.11

$$\text{Energy rate Btu/hr} = \frac{\text{Energy/cycle} \times q \times 30 \times 100 \times 60}{2 \times 12 \times 778}$$

$$= 9.64q \times \text{energy/cycle}$$

It is evident by inspection of the table that the heat transferred is equal to the heat generated when the belt thickness is equal to .007 in. approximately. Again our solution is an estimate which is conservative. The belt is not uniformly stressed and in fact at any one time only half of the belt is under tension. The heat transfer estimate is probably fairly close since the belt will assume a uniform internal temperature because different portions of it are stressed on successive revolutions since it is not running in synch with the cam. The motor is assumed to deliver constant speed independent of the loads and this leads to the belt absorbing the varying impulse loads caused by the cutter. The solution again is an order of magnitude estimate of the solution subject to refinement with better data and by testing. Again the tabular solution presented can be also done by computer and especially if data for a range of temperatures on the internal energy generation are available. The flow chart and program for this follows.

one that is relatively elastic in the sense that the length will vary depending on the driving torque, stretching under increases in torque and recovering under reduced torque. The belt must have internal material time constants short enough to do this at a rate equal to the motor speed or greater. In this way the torque surges are damped out by the belt and the excess energy of the high torque impulses is absorbed by hysteresis effects in the belt.

In several respects our design problem with the belt is somewhat simpler than with the other units in the drive train. The primary load on the belt is a tensile load used to transmit the torque from the motor shaft to the cam pulley. There is a bending force needed to move the belt around the pulley. This must be added to the tensile load, especially on the outer layer of the belt. In most cases this load is small compared with the power transmitting load and it decreases in value as the power transmitting load is transferred to the pulley by friction so that the loads are not additive.

The tensile load in the belt can be calculated from the power that the belt transmits. Using this load and the stress-strain curves, the hysteresis can be calculated. The rate of heat buildup in the belt due to the power transmission can be calculated. The impulse loading caused by the torque pulsations from the gasoline motor must be added to this load. These will be of a higher frequency than the rate of rotation of the drive pulley and will use a different stress-strain curve corresponding to the higher loading rate. The hysteresis heating effect from this is added to the effect from the power transmission. It is important to examine the dynamic response curve of the materials used. It must be determined if the material has a high degree of compliance at the loading rates used, to prevent a condition where the belt may appear to be a rigid member as a result of a resonance effect which may cause the material to react elastically.

The next stage in the design is the same as for the other two elements. The rate of heat transfer from the moving belt is calculated using an appropriate heat transfer coefficient which takes into account the air movement and the fact that the belt is also moving. A temperature is determined at which the rate of heat transfer is equal to the rate of heat generation. If no solution is possible at the initially assumed stress level, the stress level is reduced by altering the cross section of the belt and the calculation repeated until a solution is arrived at where the equilibrium temperature is below the danger level for the belt material. Since the primary stresses in this part are uniformly distributed and since the heat transfer is relatively simple, the solution for the dynamic loading on this part is the most accurate of the three elements that make up the design problem. The calculations on this part are summarized in Table 14-1.

The problem is complete except for consideration of fatigue effects and of in-use hazards and environmental exposure. The fatigue analysis consists of an examination of the S-N curves for the three materials used, to determine if the lifetime expectancy is adequate for the stress levels used. If there are no S-N data available, the only alternative is a life test on the part or S-N data accumulation for the materials. Since this application is a medium risk application involving possible personal injury, it is necessary that one route or the other be followed for good design practice.

The in-use hazards on a part like this are essentially the same as those that we have discussed with the other designs—abuse, exposure to weather, and similar problems. One aspect of the problem that must be carefully considered is exposure to elevated temperatures or malfunction of the unit that will restrict the flow of air through the unit. Since the dynamic operation of plastic parts is so dependent on the heat transfer of the frictional energy away from the part, any interference with the efficiency of this heat removal will lead to part failure.

In designing the performance and quality control tests on parts for this application, measurement of the accuracy of the parts should be considered, since the mating fit will affect the stress level. A test for part molding quality must be made to see that the part has a minimum tendency to unmold. Unmolding will be accelerated by the heat generation and any part distortion caused by the unmolding will create additional stresses, increasing the heat level and leading to premature failure. Dynamic loading applications for plastics are demanding ones and high quality is required for successful functioning of the part.

Consideration should be given to the use of thermoset materials for applications where higher temperature resistance is required. These materials have higher softening temperatures and, since they are cross-linked, higher resistance to unmolding effects. The materials frequently have higher moduli and lower hysteresis levels to make them more resistant to cyclical loading. The only significant limitations on their application is higher cost and some tendency to brittleness in the resin phase for certain materials. Canvas phenolic materials were one of the first materials to be successfully used in cams, gears, and similar applications and this type of material is in wide use in textile equipment and other industrial applications.

The analysis has covered some of the typical application types for plastics under dynamic loading. As previously indicated, the range of possibilities is quite wide, but the same basic principles apply. Essentially the analysis involves a determination of the magnitude of the dynamic loads including all of the load factors, as well as any residual static loads. The stress level allowed in the part is determined by the level that will generate heat by hysteresis at no higher level than the parts are capable of dissipating to the environment under the heat transfer conditions in the structure in which the part is used. Some of the devices which have been discussed can be added to improve the heat transfer. When equilibrium part temperatures are at safe levels the part has been properly designed, provided that the heat transfer conditions are not altered. Examination of fatigue data to determine cycles to failure is the last step in the design, and if the necessary life is possible, the design is complete. The basic steps of material selection, stress determination, and performance have been met.

Plastics make good dynamically loaded elements and are essential in many applications. The damping function is one that they perform well. In addition, plastics are lightweight and they tend to reduce the inertial loads that occur in moving parts. The fatigue levels are good and there are many applications where plastics are the most satisfactory materials in dynamic loading applications. Using this section as a guide will result in good conservative design.

Chapter 15

The Design of Plastics Parts for Electrical Applications

The early impetus for the development of modern plastics materials came from the electrical industry. To the present, one of the major areas of application for plastics is in electrical and electronic parts. The main reason for this is that plastics are inexpensive, easily shaped, dielectric materials with controllable electrical properties. There are instances where plastics materials are used in electrical applications for some specific inherent property; in most cases the plastics are used because they are good insulators.

The fact that plastics are good insulators does not mean that plastics are inert in an electrical field. They can in fact, be made to conduct electricity by the addition of fillers such as carbon black and metallic flake. The type and degree of interaction depends on the polarity of the basic resin material and the ability of an electrical field to produce ions that will cause current flows. In most applications for plastics, the intrinsic properties of the polymer are related to the performance under specific test conditions. The properties of interest are the dielectric strength, the dielectric constant at a range of frequencies, the dielectric loss factor at a range of frequencies, the volume resistivity, the surface resistivity, and the arc resistance. The last three are sensitive to moisture content in many materials. These properties are determined by the use of standardized tests described by ASTM (Table 15-1). These properties of the plastics are temperature dependent as are many of their other properties. Temperature dependance must be recognized to avoid problems in electrical products made of plastics.

The function that plastics serve in most electrical applications is that of a dielectric or insulator that separates two conductors with an electrical field between them. The field can be a steady D.C. field or an alternating field and the frequency range may vary from 0 to

Table 15-1. Electrical Test Specifications for Plastics.

Plastics Insulation for Wire and Cable

Ozone-Resistant Ethylene-Propylene Rubber Insulation for Wire and Cable, Spec. for, (**D 2802**) 33

Ozone-Resisting Ethylene-Propylene Rubber Integral Insulation and Jacket for Wire and Cable, Spec. for, (**D 2770**) 38

Polyethylene Insulated Wire and Cable, Spec. for, (**D 1351**) 38

Polyethylene Jacket for Electrical Insulated Wire and Cable, Spec. for, (**D 2308**) 38

Poly(Vinyl Chloride) Jacket for Wire and Cable, Spec. for, (**D 1047**) 38

Thermoplastic Insulated and Jacketed Wire and Cable, Testing, (**D 2633**) 38

Vinyl Chloride Plastic Insulation for Wire and Cable; 60 C Operation, Spec. for, (**D 2219**) 38

Vinyl Chloride Plastic Insulation for Wire and Cable, 75 C Operation, Spec. for, (**D 2220**) 38

Plastics, Reinforced

Glass-Reinforced Acetal Plastics for Molding and Extrusion, Spec. for, (**D 2948**) 36

Interlaminar Shear Strength of Structural Reinforced Plastics at Elevated Temperatures, Test for, (**D 2733**) 36

Tensile Properties of Oriented Fiber Composites, Test for, (**D 3039**) 33

Void Content of Reinforced Plastics, Test for, (**D 2734**) 36

Plastics, Thermal Evaluation

Heat Aging of Plastics Without Load, Rec. Practice for, (**D 3045**) 35

Thermal Evaluation of Rigid Electrical Insulating Materials (**D 2304**) 39

(Courtesy ASTM).

10^{10} Hz. The magnitude of the fields can vary from fractions of a volt such as in communications signals to millions of volts in power systems. The currents carried by the conductors range from micro-amperes to millions of amperes. With this wide range of electrical conditions the types of material that can be used must obviously be different. The selection of the materials and the configuration of the dielectric to perform under the different voltage, current, and frequency stresses is the primary design problem in electrical applications for plastics.

The primary function served by the dielectric or insulator is to separate the field-carrying conductors. This finction can be served by air or vacuum, but these media do not offer any mechanical support to the conductors. From this, the second function of the plastics insulator is derived. Since it is a mechanical support for the field-carrying conductors, the mechanical properties of the material are important. The dielectric materials interact with the electrical fields and alter the characteristics of the electrical field. In some cases this is desirable and in others it is deleterious to the operation of the system and must be minimized. This is done by both the selection of the material and the configuration of the dielectric.

To see how these concepts are applied, we will use an example of one of the major applications of plastics materials, i.e., to insulate wires, and show how a dielectric is designed to meet the service requirements. The specific requirements on a standard wire are:

1. The voltage between the conductors,
2. The current-carrying capacity.
3. The maximum operating temperature.
4. The frequency of the electric field.
5. The mechanical requirements on bending, etc.
6. Flame retardance.

The simplest wire configurement is shown in Fig. 15-1. This is a solid conductor with a sheath of insulation which might be flexible PVC or polyethylene. If the wire is rated for 600 volts power frequency A.C., the wall thickness would be about 0.020 to 0.030 inch. The dielectric strength of the PVC would be about 300 volts per mil and the polyethylene is about 600 volts per mil so that the insulation would be more than adequate at room temperature. Since the insulation value drops sharply with temperature, the wire would be limited in service temperature to 60° C, where both of these materials soften. The additional wall thickness above the theoretical minimum is used to give some mechanical strength to the insulation, to improve the resistance to cut through and bending. Since each of the conductors can handle 600 volts, it is possible to use two of the wires to handle 1200 volts. This is usually not done because of the possibility of grounding one of the conductors which would expose the other one to the full field.

The current-carrying capacity of the wire is not directly related to the dielectric. This is determined by the conductor resistance and the heating effect that it produces in the wire. The required current-

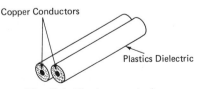

Copper Conductors

Plastics Dielectric

Fig. 15-1. Plastics coated wire.

carrying capacity determines the size of the wire and thus the size of the insulator. The temperature rise caused by the current flow determines the type of insulation to be used. If the wire is limited to 60° C service, the insulation can be one of those discussed above. If the wire is to operate at 150° C, another specification for plastic wire with better heat resistance such as polyester or fluorocarbon should be used.

If the wire is to be used to carry much higher frequency currents, the design problem in geometry and material selection becomes more complicated. A difficult high frequency insulation problem is shown in Fig. 15-2. The dielectric constant and dielectric loss values for the materials become important in the design. At a frequency of one megahertz the effect of the dielectric on the power transmission characteristics of the wire is substantial and, even at frequencies of 10 to 100 kilohertz, the insulation on the wire must be considered as a major electrical element in the circuit. When dealing with low value currents,

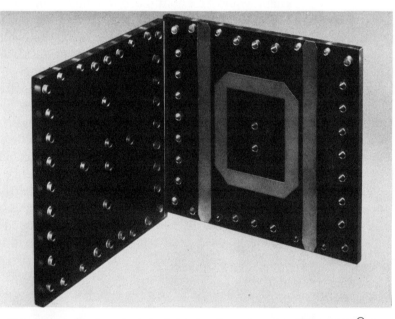

Fig. 15-2. The combined mechanical, electrical and temperature stability of PPO^(①) thermoplastic were primary factors in selecting it for this two-piece strip line used in a directional filter. Operating at temperatures from 25 to 125° F in the 1000 megahertz range, the unit is used for multiplexing frequencies. By using PPO, frequency remains nearly constant over a wide temperature range, representing an improvement ten times that of previously used materials. (*Courtesy General Electric Co.*)

the leakage resistance of the insulation is also a major problem in the application of the wire. Such wire is used primarily in communications applications and we will examine the required design modifications that the use of wires in this application requires. The leakage of current from the wire is related to the volume resistivity of the dielectric material. In most plastic materials, the volume resistivity is high and in the case of the material most used in commercial communications wire, polyethylene, the leakage is so low it causes no problems. When there is appreciable current leakage, the signal strength in the wire is reduced and noise from the environment is conducted into the wire to add to the loss of signal content (signal to noise ratio).

The value of the dielectric constant is important in the wire because of the effect that it has in coupling currents in one set of wires into another set of wires. The higher the dielectric constant, the higher the value capacitor that is formed between two wires. The capacitor thus formed is a signal carrying device at the frequencies used in communications and a signal can be capacitively coupled from one circuit to another. Polyethylene is the preferred choice for insulation of communication wire because of its low dielectric constant which minimizes the intercircuit coupling effect usually referred to as crosstalk.

The dielectric loss factor represents energy which is lost to the insulator as a result of its being subjected to alternating fields. The effect is caused by the rotation of dipoles in the polymer structure and by the displacement effects in the polymer chain caused by the electrical fields. The frictional effects cause energy absorption and the effect is analogous to the mechanical hysteresis effects except that the motion of the material is field induced instead of mechanically induced. Materials that have highly polar structures permitting the field to have strong coupling to the polymer structure have high loss factors, particularly if they exhibit large viscoelastic behavior. Materials with low polarity structures that the fields have a minor effect on, combined with a crystalline or crosslinked rigid molecular structure, show a low dielectric loss. In signal processing applications the dielectric loss represents an attenuation of the signal. Where there are large amounts of power generated, such as in a ratio transmitter, the dielectric loss represents a sufficient power drain that it will heat the material of the insulator and possibly destroy it. In both cases, it is important to minimize the amount of power dissipated into the dielectric material. This is done primarily by the use of materials which have low dielectric loss factors and generally low dielectric constants. Table 15-2 lists several materials commonly used as insulators as the dielectric constants and loss factors at several frequencies. Figure 15-3 is a curve of the dielectric loss factor as a function of frequency at two temperatures. The higher the temperature the higher the loss factor at all frequencies. This is due to the greater mobility of the polymer structure at higher temperatures permitting increased movement by the electrical fields. The greater amount of frictional energy generated by the greater excursions possible at the higher temperature increases the dielectric loss. It should be pointed out that in the case of high power applications, this tendency produces an effect similar to that in the dynamic mechanical loading in that the heating produces an increase in the ability to be heated so that, if the heat is not dissipated, the material proceeds to catastrophic destruction.

The other approach to the reduction of the power loss to the dielectric material is by reducing the amount used. This is done by replacing part of the dielectric by air, an inert gas, or by vacuum. Figure 15-4 shows three cable constructions in common use which employ

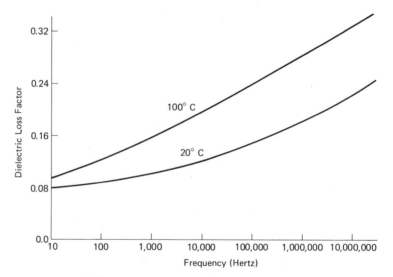

Fig. 15-3. Frequency vs dielectric loss factor.

Table 15-2. Electrical Properties of Plastics.

Material	Resistivity		Dielectric Constant/Dissipation Factor					
	Volume	Surface	100 Hz	1 kHz	1 mHz	10 mHz	100 mHz	1000 mHz
ABS	2×10^{16}	10^{14}	.005/2.9	.006/2.8	.008/2.8	.007/2.8	.005/2.7	.001/2.7
Acrylic	10^{18}	10^{14}	.062/3.6	.058/3.2	.045/3.1	.033/2.9		
Cellulose Ester	3×10^{15}	10^{14}	.006/3.8	.011/3.6	.024/3.3	.022/3.2	.020/3.0	.014/2.1
FEP	10^{18}	10^{16}	.0005/2.1	.0005/2.1	.0005/2.1	.0005/2.1	.0008/2.09	.0007/2.05
Nylon 6	10^{15}	10^{13}	.031/4.2	.024/3.8	.031/3.8	.020/4.0		
Polycarbonate	10^{16}	10^{15}	.001/3.1	.0013/3.1	.007/3.1	.011/3.1	.015/3.1	
Polyethylene	10^{19}	10^{16}	.0001/2.34	.0001/2.34	.0001/2.34	.0001/2.34	.0001/2.34	.0001/2.34
Alkyd	10^{13}	10^{14}	.02/6.0	.02/5.8	.015/5.4			
DAP (SD15)	10^{16}	10^{13}	.026/3.8	.020/3.7	.016/3.6			
Phenolic MFE	10^{14}	10^{9}	.013/5.4	.013/5.3	.033/4.9			
Epoxy	10^{16}	10^{14}	.004/3.22	.004/3.25	.004/3.25			

March/April 1976 Plastics Design Forum, 85.

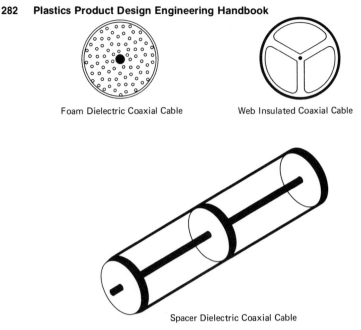

Foam Dielectric Coaxial Cable   Web Insulated Coaxial Cable

Spacer Dielectric Coaxial Cable

Fig. 15-4. Low loss cable configurations.

these approaches to minimize dielectric loss. The first is the use of a foamed dielectric material which is commonly used in either twin lead transmission lines or in coaxial cables used for antenna lead-in wires in the UHF-TV antenna applications. The second system, which is illustrative of several sectional spacers, is used widely in communications cables of the coaxial type to minimize losses to the dielectric by reducing the amount of dielectric material in the cable. The cable design must be modified to take into account the lower dielectric constant of the air which tends to increase the diameter of the cable so that it is not a simple replacement situation. The additional diameter will tend to increase the amount of material required so that an optimum must be reached in terms of the geometry to reduce the material to a minimum and still have a mechanically stable cable structure. The third scheme uses bead-like spacers at intervals along the cable. This type of cable is frequently evacuated to improve the dielectric performance of the cable.

The second major area for the use of plastics in electrical applications is at the terminations of the conductors. The connectors which are used to tie the wires into the equipment using the power, or used to connect the wires to the power source, are rigid members with spaced contacts. These are designed to connect with a mating unit and to the extension wires. The other type of wire termination is a terminal board where there are means to secure the ends of the wire leading to the equipment and the internal wiring in the equipment. Examples of these are shown in Figs. 15-5, 15-6, and 15-7. These termination units require: (1) adequate dielectric strength to resist the electric field between the conductors, (2) good surface resistivity to prevent leakage of current across the surface of the material of the connector, (3) good arc resistance to prevent permanent damage to the surface of the unit in case of an accidental arc over, and (4) good mechanical properties to permit accurate alignment of the connector elements so that the connectors can be mated properly. If the connectors are to be used in high frequency applications, they must be made of materials with low dielectric loss to avoid either damage to the part or

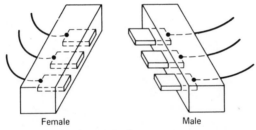

Female   Male

Fig. 15-5. Connector.

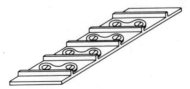

Fig. 15-6. Terminal board.

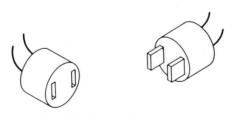

Fig. 15-7. Household 2-prong plug.

signal loss in the circuit. Common commercial connectors are depicted in Figs. 15-8 and 15-9.

The design of a connector is a fairly straightforward process. It is easily illustrated with the example of a two-element connector of the type that may be used for either a power connection or to plug in an audio system. Figure 15-7 depicts a typical unit. The material selected will be one that has the required dielectric strength at the maximum operating temperatures and at the frequency of intended use. Table 15-3 lists several materials that are widely used as connector materials and typical types of connectors in which they are employed. For a power connector phenolic, polyester glass, or semi-rigid PVC might be used. The anticipated voltage is used to calculate the probable leakage current which should be less that 0.1% of the magnitude of the current in the conductor. The arc resistance of the material would not be critical for an appliance plug, since this condition rarely occurs

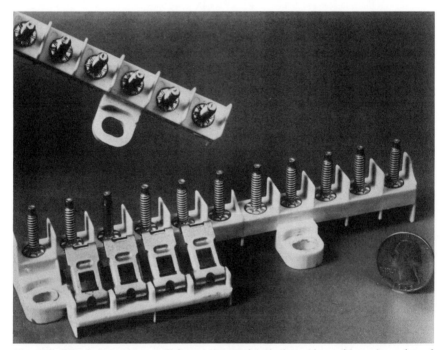

Fig. 15-8. Electrical connectors call for tough materials that are able to withstand a variety of environments and that are capable of being molded economically to exacting standards. Electrical connectors such as these molded from thermoplastic polyester are found in dozens of applications ranging from appliances and automobiles to oil drilling rigs and television sets.

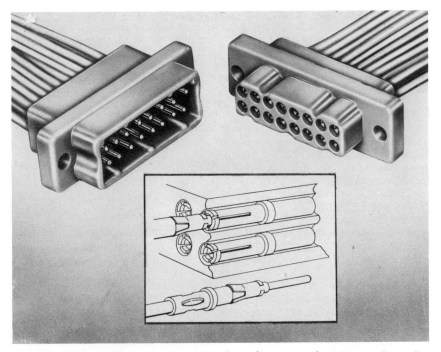

Fig. 15-9. This general-purpose connector employs crimp-on-snap-in contacts and one-piece insulator sections molded of polycarbonate.

in this application. For connectors used for industrial power connections, the material should have good arc resistance because of the possibility of flashover. Total arc resistance is provided by inorganic materials such as glass-bonded mica. Depicted in Fig. 15-10 is a power plant lightning arrestor with glass-bonded mica for total arc resistance. The remainder of the problem would be to make sure that the connector is stiff enough to hold the contact members in alignment when the connector is inserted into the mating receptacle. The receptacle design is essentially the same as the plug except that it has the opposite set of contacting elements.

When the connector is used to make connection in an audio circuit, the configuration can be essentially the same. The additional consideration is that the material have low loss dielectric at the frequencies to be transmitted and that the spacing of the contact elements be determined by the transmission line characteristics required. Spacing is an electrical design function and its determination re-

Table 15-3. Surface Resistivity of Various Plastics.

Material	Equilibrium Surface Resistivity at 100% RH	% RH to Produce a 10-Fold Change in Surface Resistivity	Recovery Time in Minutes After Exposure at 100 % RH for 1 hr	16 hr
Cellulose acetate butyrate	$>2.0 \times 10^{13}$	—	0	0
Silicone rubber	1.0×10^{13}	—	0.13	—
Polytetrafluoroethylene	3.6×10^{12}	—	0.17	0.17
Polystyrene (sheet)	8.4×10^{11}	—	0.13	0.13
Polydichlorostyrene 2-5	2.9×10^{10}	7	0.17	—
Ethyl cellulose	1.3×10^{10}	9	0.33	0.5
Cellulose acetate	7.0×10^{9}	6	1.0	6
Polyvinyl chloride-acetate	5.7×10^{9}	12	6	—
Molded phenolic, mica-filled	5.0×10^{9}	9	0.17	13
Polyamide (nylon)	3.8×10^{9}	14	200	—
Polystyrene (molded)	2.4×10^{9}	10	0.17	0.17
Polyethylene	1.3×10^{9}	9	0.17	0.17
Paper-phenolic laminate, XX	1.3×10^{9}	16	80	—
Molded phenolic, asbestos-filled	1.2×10^{9}	9	1.5	100
Paper-phenolic laminate XXXP	6.6×10^{8}	15	300	—
Cotton cloth-phenolic laminate, LE	5.0×10^{8}	18	400	—
Molded phenolic, mica-filled	3.2×10^{8}	8	40	—
Polydichlorostyrene 3-4	2.4×10^{8}	6	0.33	5.3
Molded phenolic, cellulose-filled	2.3×10^{8}	10	400	—
Glass mat-analine formaldehyde laminate	2.4×10^{8}	9	14	1000
Cotton cloth-phenolic laminate, C	2.2×10^{8}	16	300	—
Vulcanized fiber	2.2×10^{8}	—	6000	—
Glass cloth-melamine laminate	3.8×10^{7}	14	300	—
Molded phenolic, mica-filled	3.0×10^{7}	11	7	—

Courtesy *Engineering Design for Plastics*, Eric Baer. New York, Van Nostrand Reinhold Co., 1964.

quires knowledge of the desired transmission line characteristics of the circuit. This part of the design is usually done by the electrical engineer and is an operating parameter for the plastics designer. The voltage resistance and other design factors are based on the data usually supplied for the material by the manufacturer.

In many electronic and electrical applications the internal wiring of the systems is done by "printed" methods as depicted in Fig. 15-11. The substrates on which the printed wiring is done are usually plastics. Commonly used materials are paper-based phenolic laminates,

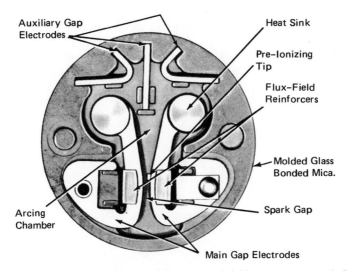

Auxiliary Gap
Electrodes

Heat Sink

Pre-Ionizing
Tip

Flux-Field
Reinforcers

Molded Glass
Bonded Mica.

Spark Gap

Arcing
Chamber

Main Gap Electrodes

Fig. 15-10. As the gap sparks over, power-follow current is initiated and a magnetic field is established, setting up magnetic flux in the gap area. Flux-field reinforcers, with a higher permeability than air, lower the reluctance of the flux path, causing an increase in flux density in the arc path. Reinforcement of the field behind the arc causes the follow-current arc to move rapidly away from the spark gap toward the auxiliary electrodes. (*Courtesy Mykroy Ceramics Co.*)

Mylar® and epoxy glass laminates. These internal wiring assemblies introduce a new design area in this application because of basically good electrical properties and generally good chemical resistance to the chemicals and solvents used in processing the printed circuitry. They cannot be used in applications for high frequency circuitry because of the dielectric loss of the two materials. The phenolic is poor even at low RF frequencies and the epoxy resins have high loss factors at higher RF frequencies. For these applications the printed circuit board materials used have been polyester glass, silicone glass, teflon glass, glass-bonded mica, and polyolefin-glass. There has been recent interest in the thermoplastic materials for printed circuit base materials. Materials such as glass-filled polyester have good electrical and mechanical properties and with the contemporary wave soldering techniques it is possible to solder the boards without distortion.

The introduction of thermoplastic materials into the area of printed circuit substrates has led to a broader type of application for the circuits. The parts can be molded and the circuit applied to a molded part which will have a molded-in connector structure used to inter-

Fig. 15-11. Copper-clad reinforced epoxy and copper-clad Mylar circuiting are depicted with circuitry. (*Courtesy Westinghouse*)

connect the device to the rest of the system. By combining the connector and substrate functions, it is possible to make very compact printed circuit units. Such units are used in the new watches with electronic drives instead of the traditional mechanical spring driven drives. Another area where the combination function is being explored is in large-scale integrated circuit unit supports where the complex interconnection requirements make a combination circuit support and printed circuit unit an attractive way to achieve high packing efficiency.

When using molded plastics parts such as the connectors and the circuit supports, it is important to make sure that the moldings, in addition to being made to close tolerances, be made under molding conditions that make for stable parts. Electrical parts are subjected to strong electrical fields in addition to the usual environmental abuse. Distortion of the part can lead to serious electrical malfunction by changing spacings which will alter electrical characteristics and if extreme enough, result in short circuits with serious results. Designs for electrical parts made of plastics should take into account the effect of the processing on the part performance. In addition to possible distortion from heat, improper molding conditions can lead to premature failure from the effect of chemical agents if the part is used in an etched circuit application or from the effect of corona degradation if any part is used in a high voltage application. Conductor-to-insulator must be tight to eliminate corona.

There are two areas of application for electrical insulation where the effects of long time exposure to electrical fields produces a fatigue effect that can be compared to the creep that occurs under static stress or the fatigue effects that occur under sustained dynamic loading. The first effect is encountered in high voltage D.C. cables such as are used in X-ray equipment, some industrial and research equipment, and in the proposed new high voltage D.C. distribution systems. Under sustained D.C. fields the plastic moves internally so that the dipoles align themselves in the field in response to the continuing voltage stress. As this continues, the dielectric constant of the material increases and the dipoles begin to break loose and migrate through the material. In doing so they disrupt the structure of the material, reducing the dielectric strength. After an extended period of time, usually several years, the dielectric will spontaneously break down and the system will arc over. This effect is similar to mechanical creep since the same sort of field based diffusion effects are at work to produce structural changes that take place.

The second effect is caused by the operation of alternating electrical fields on the dielectric in the system. This can happen to insulation at power frequencies as well as higher frequencies (Fig. 15-12). This is not the dielectric heating effect mentioned earlier, but an actual disruption of the polymer structure caused by the alternating stresses imposed by the alternating field. Some regions of the structure develop higher than average stresses and the polymer structure is broken at these points. As the number of defects grow with time, the structure becomes electrically less resistant to the imposed fields and the

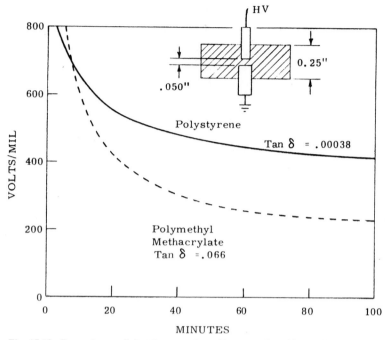

Fig. 15-12. Dependence of electric strength at 60 cps on time (thermal breakdown).

dielectric strength decreases to the point where the field arcs over. These effects occur after a period of years, usually in electrical insulation that is operating near the limit of its dielectric strength. The same type of S-N analysis that is used in mechanical fatigue is used in predicting this type of electrical fatigue. It is essential that any materials used for this type of service be carefully evaluated for fatigue and arc resistant to this effect.

In terms of environmental exposure, water and humidity must be carefully evaluated in electrical applications. In general, if a plastics material absorbs a significant amount of water, the electrical resistivity drops. This is the case for nylons and for phenolic as can be seen from Table 15-3. Care must be used in selecting a dielectric to insure that the electrical properties such as the insulation resistance and dielectric strength, as well as other electrical properties are adequate under the conditions of use, particularly if this involves exposure to high humidity conditions. Temperature causes changes in most electrical properties (see Figs. 15-13, 15-14, 15-15, 15-16).

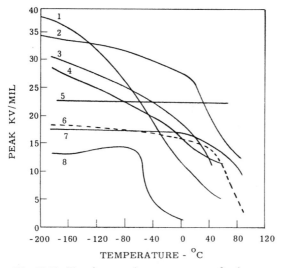

Fig. 15-13. Electric strength *vs* temperature of polymers.

1—Polyvinyl alcohol
2—Polymethylmethacrylate
3—Polyvinylchloracetate
4—Chlorinated polyethylene
5—Mica
6—Polystyrene
7—Polyethylene
8—Polyisobutylene

There is another type of condition that results from exposure to high humidity. The alteration in electrical properties caused by moisture absorption in nylon and phenolics is reversible. When the moisture content is decreased, the properties of the materials recover to close to the original values. In some instances the exposure to moisture and electrical fields can cause irreversible damage that can lead to failure. One case is that of the polyester materials such as the alkyd molding compounds. These materials, when exposed to continuous high humidity, especially in the presence of an electrical field, hydrolyze into the acid and alcohol precursors from which they are made. The acid plus water present make a conductive material that will cause the material to short the electrical circuit. The process by which the decomposition of the polyester takes place is very gradual at first and then accelerates so that extended testing of the material is necessary to be sure that the particular polyester composition used is resistant to hydrolytic degradation.

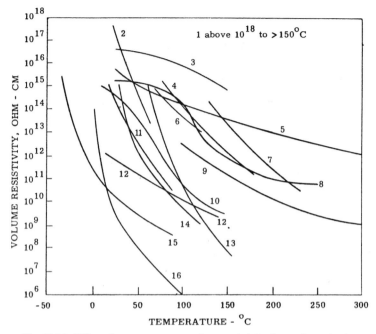

Fig. 15-14. Effect of temperature on volume resistivity for various plastics.

1—Polytetrafluoroethylene (Dupont "Teflon")
2—Low-density polyethylene
3—Polycarbonate foil (Bayer "Makrafol N")
4—Polyester-glass fabric laminate
5—Solventless silicone resin
6—Epoxy casting resin ("Furane Epocast #3")
7—Polyethylene terephthalate (Dupont "Mylar")
8—Silicone rubber
9—Molded silicone resin
10—Plasticized cellulose acetate butyrate foil (Bayer "Triafol, BW")
11—Plasticized vinyl chloride
12—Soft polyurethane foam (Bayer "Moltoprene"—density 30 kg/m³)
13—Epoxy resin
14—Cast epoxy resin (Ciba "Araldite 502")
15—Polyurethane elastomer (Bayer "Vulkollan 18")
16—Polyamide resin (nylon)

One other long-term condition that takes place with relatively low level D.C. fields in the presence of moisture is the migration of the metal of the conductor into the plastic. This has been discovered to be a common thing with silver conductors and phenolic insulators. The first instance of field failures were discovered in telephone equipment. The problem can occur with other metals with phenolic and also conceivably with other plastics that are moisture sensitive and can have a solvating action on the conductor metals that they are in con-

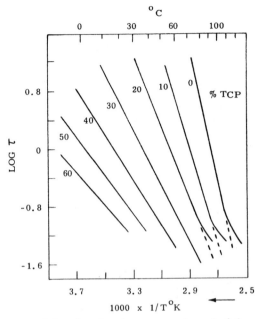

Log of the relaxation time, τ, vs reciprocal of the absolute temperature; polyvinyl chloride-tricresyl phosphate system.

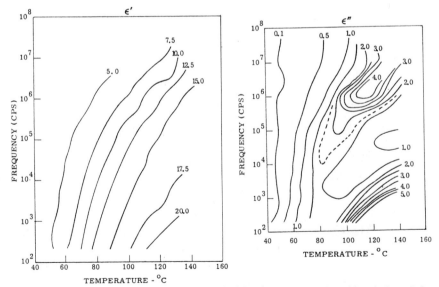

Fig. 15-15. Frequency-temperature contours of dielectric constants, ϵ', and loss index, ϵ'', for nylon.

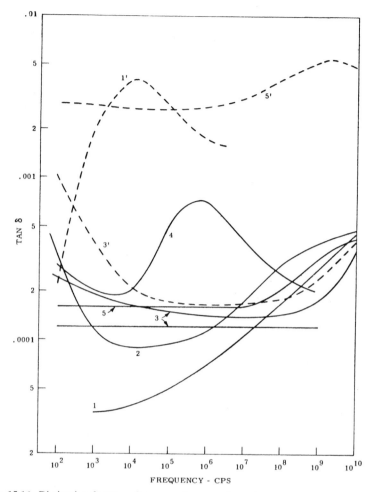

Fig. 15-16. Dissipation factor *vs* frequency for nonpolar polymers.
1—Polystyrene at 25°C, 1'—Polystyrene at 134.5°C
2—Polyisobutylene at 25°C
3—Polytetrafluoroethylene at 23°C (two sources); 3'—Polytetrafluoroethylene at 100°C
4—Polyfluoroethylene-propylene at 23°C (Dupont FEP)
5—Polyethylene at 25°C; 5'—Polyethylene at 25°C after milling at 190°C for 30 minutes.

tact with. Most plastic materials should be avoided inside hermetically sealed containers with movable contacts. Vapors released from the organic plastics deposit on the contacts to produce an insulation layer leading to contact failure.

There are several applications for plastics materials in electrical devices which use the intrinsic characteristics of the plastics material for the effect on the electrical circuit. The most obvious of these is the use of plastics, particularly in the form of thin films as the dielectric in capacitors. Polyester films such as Mylar (Fig. 15-17) are especially useful for this type of application because of the high dielectric strength in conjunction with a good dielectric constant. Mylar has the additional desirable feature that it is available in very thin films down to 2.5 microns. Since the value of a capacitor is directly proportional to the area and inversely proportional to the spacing of the conductive plates, the thinner materials permit high values of capacitance in small size units. There are other materials that make good capacitors such as polyvinylidene fluoride which has a very high dielectric constant and good dielectric strength and oriented polystyrene which makes a good capacitor for high frequencies because of its low dielectric loss constant.

Another application for plastics which uses the intrinsic properties is in electrets. Some materials such as highly polar plastics can be cooled from the melt under an intense electrical field and develop a permanent electrical field which is constantly on or constantly renewable. These electret materials are finding a wide range of applications that vary from uses in electrostatic printing processes, to supplying static fields for electronic devices, to some specialized medical applications where it has been found that the field inhibits clotting in vivo. One application for the electret material is in a microphone which has a high degree of sensitivity and the electrical waves are produced by the field variations caused by the change in spacing of an electrode to an electret.

A wide range of applications in electronics makes use of the plastics as a structural binder to hold active materials. For example, a resin such as polyvinylidene fluoride is filled with an electroluminescent phosphor to form the dielectric element in electroluminescent lamps. Plastic resins are loaded with barium titanate and other high dielectric powders to make slugs for high K capacitors. The cores in high frequency transformers are made using iron and iron oxide powders bonded with a resin and molded to form the magnetic core. Magnetic recording tape, in addition to using plastic films as a support for the recording surface, also use polyvinyl alcohol and urethane resins as binders for the magnetic oxides that form the recording medium. The range of special characteristic materials that can be made from plastics is broad.

A new field emerging in electronics is in the application of mate-

Temp. (°C)	60 Cycles per Second Dielectric Constant	60 Cycles per Second Dissipation Factor (%)	1 Kilocycle per Second Dielectric Constant	1 Kilocycle per Second Dissipation Factor (%)	1 Megacycle per Second Dielectric Constant	1 Megacycle per Second Dissipation Factor (%)
0	3.02	0.5	2.99	0.80	2.79	1.6
25	3.02	0.21	3.01	0.47	2.83	1.9
75	3.02	0.14	2.99	0.12	2.95	1.7
100	3.08	0.69	3.06	0.46	3.01	1.4
125	3.36	1.2	3.29	1.4	3.15	1.7
150	3.46	0.64	3.45	0.68	3.31	2.0

ASTM D 150-47T—Film pile up.

Volume Resistivity at 500 Volts D.C.

°C	Ohm-Centimeters
0	Greater than 2.1×10^{15}
25	" " 2.1×10^{15}
75	" " 2.1×10^{15}
100	" " 2.1×10^{15}
125	" " 7.4×10^{14}
150	" " 6.9×10^{13}

Dielectric Strength at 60 cps
(Between 2″ Disc Electrodes in Air)

°C	Volts Per Mil
0	4400
25	4500
75	4200
100	4300
125	3250
150	3150

ASTM D 149-44—Short Time Method.

Film Properties vs Temperature

°C	Tensile Strength (psi)	Elongation (%)
-18	26,000	135
25	26,000	130
50	10,200	71
100	6,800	68
138	4,500	63
150	4,000	62
160	3,300	62

Fig. 15-17. Electrical properties of DuPont Mylar, a polyester film.

rials which are classified as liquid crystals. These materials exhibit some of the properties of crystalline solids and still flow easily like liquids. One group of these materials is based on low polymer materials with strong field interacting side chains. Using these materials, there has developed a new field of electro-optic devices whose characteristics can be changed sharply by the application of an electric field. These will be discussed in more detail in the chapter on optical design.

A number of areas in which plastics are used in electrical and electronic design have been covered. The basic use for plastics is as an insulator or as a dielectric material and the problem of design is to electrically separate two conductors having an electrical field between them. The effect of field intensity, frequency, environmental effects, temperature, and time were reviewed as part of the design process. Several special applications for plastics based on intrinsic properties of plastics materials were discussed.

The use of plastics in electrical applications is one of the oldest and most important. Other areas such as static electricity and its use and control were not discussed since they represent a different type of application. As new materials become available and the electrical arts develop, the uses for plastics in electrical applications will increase, both in the basic application as a dielectric and in special applications using the special intrinsic properties of the polymer materials.

Chapter 16

Design of Plastics Parts for Optical Applications

Many plastics materials are transparent and most are translucent in the unpigmented state. They have a range of optical properties that make them interesting for a wide spectrum of optical applications that extends from windows to lens systems to sophisticated applications involving action on polarized light. This chapter will cover some of the applications for plastics in optical devices and the design principles involved.

The application for plastics most widely known, based on the transparency and clarity of plastics, is their use as a window or cover. There are a number of inexpensive plastics such as polystyrene, the cellulosics, the vinyls, and the acrylics that have been widely used to make boxes for displaying merchandise, for windows in instruments, and for glazing applications for buildings. The primary optical property used in these applications is the transparency and lack of haze or light scatter. Table 16-1 lists the transparent plastics which are commercially available. The primary requisite for these materials is their physical and aging properties since the other optical properties are not of importance. Design of a box, a light, or a display unit will involve the requirements for static and/or dynamic loading that may be encountered in the end-use application. The only change is that the part is designed with a clear transparent material.

One of the applications for plastics in optics is in refracting and reflecting elements where they can be used as glass in lenses, prisms, mirror supports, and other refracting and reflecting units. The range of refractive index for plastics is generally in the same range as that for optical glasses. (Refractive index and other optical terms are defined in Table 16-2.) As a result, lenses having the same general properties as glass can be made from plastics such as the acrylics and polystyrene. There are two major differences between glass lenses and

Table 16-1. Properties of Some Optical Plastics.

Properties	ASTM Method	Units	Methyl Methacrylate (Acrylic)	Polystyrene (Styrene)	Polycarbonate	Methyl Methacrylate Styrene Copolymer
Refractive index (n_D)	D 542		1.491	1.590	1.586	1.562
Abbe. value (v)	D 542		57.2	30.9	34.7	35
$dn/dt \times 10^{-5}/°C$			8.5	12.0	14.3	14.0
Haze (%)	D 1003	%	<2	<3	<3.0	<3
Luminous transmittance (0.125-in. thickness)	D 1003	%	92	88	89	90
Critical angle (i_c)	D 648-56	degree	42.2	39.0	39.1	39.6
Deflection temperature		°F				
3.6 F/min, 264 psi			198	180	280	212
3.6F/min, 66 psi			214	230	270	
Coefficient of linear thermal expansion	D 696-44	in./in./°F × 10^{-5}	3.6	3.5	3.8	3.6
Recommended max. cont. service temp.		°F	198	180	255	200
Water absorption (Immersed 24 hr at 73°F)	D 570-63	%	0.3	0.2	0.15	0.15
Specific gravity (density)	D 792		1.19	1.06	1.20	1.09
Hardness (0.25-in. sample)	D 785-62		M 97	M 90	M 70	M 75
Impact strength (Izod Notch)	D 256	ft-lb/in.	0.3–0.5	0.35	12–17	3.6
Dielectric strength	D 149-64	V/mil	500	500	400	450
Dielectric constant:						
60 HZ	D 150		3.7	2.6	2.90	3.40
10^6 Hz			2.2	2.45	2.88	2.90
Power factor:						
60 Hz	D 150		0.05	0.0002	0.0007	0.006
10^6 Hz			0.03	0.0002–0.0004	0.0075	0.013
Volume resistivity	D 257	ohm-cm	10^{18}	$>10^{16}$	8×10^{16}	10^{15}
Trade names			LUCITE PLEXIGLAS	DYLENE STYRON	LEXAN MERLON	NAS

*This information is taken from raw material manufacturers' available published data. Specific material formulation data should be confirmed prior to design and specification.
(*Courtesy U.S. Precision Lens, Inc.*)

Table 16-2.

Refractive index	The ratio of the velocity of light in free space to the velocity of light in the medium
Light scattering	The change in direction of a portion of the light transmitted due to refraction or reflection at the surfaces of inclusions in the material
Birefringence	The property of anisotropic optical media which causes polarized light with one orientation to travel with a different velocity than polarized light with another orientation
Polarized light	Light which has the electric field vector of all of the energy vibrating in the same plane. Looking into the end of a beam of polarized one would see the electric field vectors as parallel or coincident lines
Dichroism	A property of an optical material which causes light of some wavelengths to be absorbed when the incident light has its electric field vector in a particular orientation and not absorbed when the electric field vector has other orientations.
Light transmissability	The ratio of the light exiting from an optical material to the light entering the material
Haze	The cloudy appearance in a plastic material caused by inclusions which produce light scattering
Color	The sum effect of the wavelengths of light transmitted by or reflected from a material
Dispersion	A property of an optical material which causes some wavelengths of light to be transmitted through the material at different velocities and the velocity is a function of the wavelength. This causes each wavelength of light to have a different refractive index

plastic lenses that must be considered from the standpoint of use. The first is that the plastics have a much greater change of dimension with temperature and a much greater change of optical constants with temperature. The other major difference is that, while the plastics are much more resistant to impact than glass, their resistance to scratching and to deformation is much lower than that of glass. As a result of these limitations, most of the applications for plastics in optical elements are for low precision optics. It would be very difficult to use plastics materials for the parts of a compound microscope to be used under normal conditions.

One of the major advantages of plastics optical elements is that they can me made by molding or casting to good accuracy at low cost. Some glass lenses are pressed from hot glass but the majority of the lenses are ground and polished from rough blanks. Glass lenses are not molded because it is difficult to get good surfaces and because the strains left in the glass as a result of the pressing operation affect the optical properties of the lenses. In the case of plastics it is possible to get excellent surface quality using precision molds and ap-

propriate molding methods. The plastics may also have strains generated in the materials as well as flow orientation caused by the molding
process. In the case of plastics, however, the effect on the usability
of simple lens systems is small. Consequently, there are a large number of magnifiers, simple lenses, prisms, and opthalmic lenses molded
from plastics. They have the advantage of being resistant to breakage and, properly handled, are much more likely to survive rough
usage. Lenses for safety glasses are made from highly impact resistant
plastics such as modified acrylics, ADC, and polycarbonate. They
will resist puncture from flying objects and offer the exceptional
eye protection. They can be molded or ground and polished to pre-
scription requirements. Allyl diglycol carbonate (ADC) is the most
highly scratch resistant of the transparent plastics; it is a thermosetting material.

The ophthalmic applications for plastics lenses include contact
lenses which are now made of acrylic plastics. Another material for
this application is a special hydrophillic acrylic polymer used in soft
contact lenses. These lenses are claimed to be much more comfortable
than rigid contact lenses.

The design of optical elements from plastics follows conventional
optical design procedures which are covered in many texts on lens and
prism design. The basic principles of design are based on ray tracing
to determine the focal point of a lens or the image distance for a lens
system. Figure 16-1 shows how the refractive index is used to make a
ray trace through a transparent optical material using the refractive
index of the material. For the geometry of a suitable optical element
it is necessary to use a text on optical design such as the *Handbook of
Plastic Optics* by U.S. Precision Lens Inc. or the services of an optical
designer. The engineering aspects of the design are first, selection of
the appropriate material and then, the design of the part and process
to make an accurate strain free part.

In addition to the differences mentioned above, the plastics are
different in another optical property from optical glass or crystalline
optical materials. The degree of dispersion of light is much greater
for plastics than for glass. Dispersion is the difference in refractive
index for the different wavelengths of light and it is greater for plastics.
As a result, a plastics prism would separate the different colors of the
spectrum much more than a glass prism would. This characteristic
makes it more difficult to make lenses of plastics without fringing
colors.

Index of Refraction $= \dfrac{\text{Sin } i}{\text{Sin } r}$ Air Into Refractive Medium

$\dfrac{1}{\text{Index of Refraction}} = \dfrac{\text{Sin } i}{\text{Sin } r}$ Refractive Medium into Air

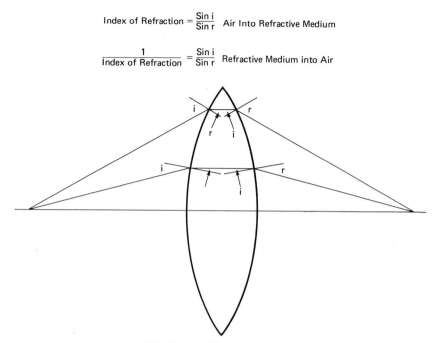

Fig. 16-1. Refraction in a lens.

colors. Despite this limitation, good camera lenses corrected enough to use for color photography have been made out of plastics for low cost cameras. It is possible to design around the limitations of the plastics materials when the cost advantage justifies the additional design effort.

To be successful, molded optical elements of plastics must be produced with careful control of the molding process. In the case of optical parts it is particularly important that the molding conditions be carefully controlled to minimize molded-in strains. In addition to these strains reducing the dimensional stability of the parts leading to distorted images, the strains themselves affect the quality of the image. This is a result of the fact that the strained areas have a different refractive index from that of the unstrained areas with resultant distortion of the image. Most optical parts are molded in special molds which sustain pressure on the part as it cools to give good surface quality. In addition, the controls on the molding machines are of the best type to insure close control over the melt temperature and pressure. Static mixers are employed to eliminate thermal gradients in the material. Schemes for quality control use the optical image quality as a test, and use polarized light inspection methods to check for residual molding strains and orientation.

The high degree of clarity and low haze of some plastics, particularly methyl methacrylate, makes possible the use of these materials in applications using the light piping effect. The physical optics of this effect are shown in Figure 16-2. Any light that enters the end of a long rod of a clear transparent refractive medium at angles less than the critical angle (defined as the angle at which light is refracted parallel to the surface) is trapped in the rod and transmitted down the length of the rod by multiple internal reflection. The trapped light is re-emitted at the end of the rod or plate, or at any place along the length of the "light pipe," where the surface is changed in angle or where the surface is roughened to form a light diffusing area. Using materials such as methyl methacrylate which are very clear, have very little light absorbtion in the visible spectrum, and have a very low haze level to scatter the light and change direction, the light can be piped over distances of the order of three to four meters with a minimum of light attenuation.

This effect is used in several areas of plastics product design. One application is in the illumination of instrument dials and similar indicia where it is impractical to use lamps close to the indicia. The lamps can be placed in a convenient location and the light piped to the indicia where the surfaces are shaped to release the light (Fig. 16-3). Another application is for lights which must be inserted in confined spaces where suitable bright light sources either do not fit or are too hot to use. An example of such an application is the light used in medical practice for examination of a patient's throat (Fig. 16-4).

The effect is widely used in signs and display devices to make them self illuminated. Edge lighted signs and panels are widely used in aircraft and for general display and office use. The basic design of signs

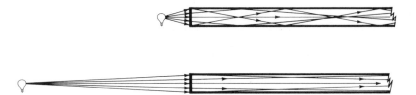

Fig. 16-2. Effect of bulb location on collimation of light when piping light through transparent acrylic rod. (*Courtesy E. I. DuPont*)

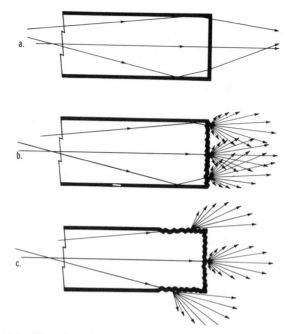

Fig. 16-3. Effect of roughened output end surface on distribution of light.

is shown in Fig. 16-5. A sheet of acrylic material has light introduced into one edge from suitable lamps. The light is carried across the sheet. The indicia that are to be displayed are cut into the surface opposite the side from which the sign is to be viewed. The indicia can be either polished angle cuts or an area which is roughened to be a diffusing surface. In either case, the light piped through the sheet is altered in direction, emitted from the sheet toward the viewer, and appears as a self-illuminated sign.

This effect is also used in the transmission of light to form images as well as to transmit small areas of light. This is the application in fiber optics. An optic fiber consists of a small diameter monofilament of a clear plastic such as methyl methacrylate which has a thin coaxial layer of another clear material of lower refractive index. The mono-filaments are of the order to 10 to 50 microns in diameter and the coating is usually 5 to 10% of the diameter. These fibers are very efficient light piping elements as a result of the coatings. A bundle of optic fibers can transmit light over distances of 10 to 20 meters with a

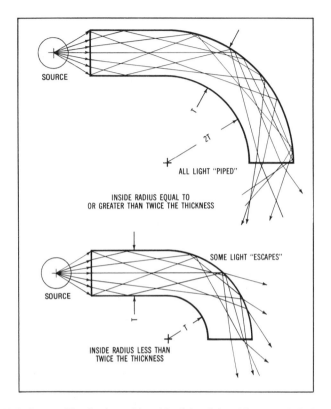

SOURCE

T

2T

ALL LIGHT "PIPED"

INSIDE RADIUS EQUAL TO
OR GREATER THAN TWICE THE THICKNESS

SOME LIGHT "ESCAPES"

SOURCE

T

T

INSIDE RADIUS LESS THAN
TWICE THE THICKNESS

Fig. 16-4. Remote illumination achieved by light piping. (*Courtesy E. I. DuPont*)

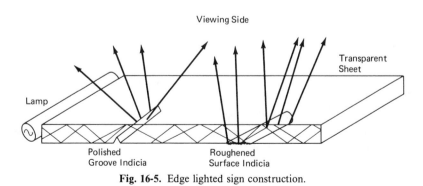

Viewing Side

Transparent
Sheet

Lamp

Polished
Groove Indicia

Roughened
Surface Indicia

Fig. 16-5. Edge lighted sign construction.

low degree of attenuation. Random bundles of fiber optics make a good medium for bringing light to a specific region for illumination. A coherent bundle of fibers (one in which the fibers are aligned so that they occupy the same position everywhere along the length of the bundle) can be used to transmit images over long distances, around corners, and past other obstacles. The image can be viewed directly by looking into the end of the bundle. When the object is properly illuminated it can be seen directly. In other cases the object is imaged on one end of the fiber optic bundle and observed at the other end as a clear image as if it were the focal plane of the lens. This feature of the fiber optics is used in a number of optical systems for remote viewing. One major application is for cyctoscopes in medical diagnosis and for borescopes used to examine inaccessible areas in machinery. Coherent bundles of fibers, properly transposed, are used as an encoding and decoding device to handle confidential image information. Short bundles of the fibers are used in conjunction with cathode ray tubes and other self-illuminated displays to improve visual contrast by minimizing the effect of ambient light on the display. Other applications for fiber optics are for decorative effects and for optical and photographic applications where the ability to transmit an image or to alter it in a predescribed manner simplifies the system.

In using fiber optics the designer is mainly concerned with a standardized material which has specific characteristics in terms of optical performance. Fiber optics made of plastics can be affected by exposure to the environment with deterioration of performance. Heat is an important environmental factor and the most likely cause of damage in optical applications. The heat is generated by the light sources used. Some of the infrared generated by light sources can be removed with the use of appropriate filters and, in cases where intense hot light sources are used, this is good design practice.

Plastics are the preferred optical material used in lenses for controlling the light in warning lamps such as emergency lights, stop lights on cars, and the retroreflective lenses used on cars and on highway signs to show the presence of an obstacle. These lenses are usually made with a large number of specially shaped lenticulations which are used to direct the incident or transmitted light in a direction where it can be readily seen. A typical lens of this type is shown in Fig. 16-6. The lenticulations may be pyramidal or they may be spherical sections. Both can be designed as excellent light directors. The taillight

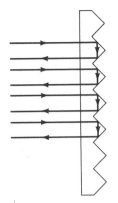

Fig. 16-6. Retroreflecting lens.

lens on automobiles probably represents one of the largest applications of optical plastics and the retroreflector element used both on cars and highway posts is probably another major use of optical plastics.

Another group of optical elements use plastics in very fine patterns to make special optical elements. One of these is the Fresnel lens which is a collapsed lens structure that has the effect of a strong magnifier but is essentially a flat sheet. This unit is shown in Fig. 16-7. The lens is made by a special molding technique from a carefully machined master. It is used as a focusing lens for light sources, as an intensity leveling unit in reflex camera viewers, and as a coarse view magnifier of simple objects. A "fish-eye" viewer is shown in Fig. 16-8. Large Fresnel lenses are used in solar furnaces to gather large areas of sunlight and to focus it at a point to achieve high temperatures. The other fine pattern application for plastics is in replica diffraction gratings. When a pattern of lines is made with a count of 50 to 500 lines per millimeter it acts as a diffraction grating which can break light into its components by selective interference. Diffraction gratings are made by ruling lines on a metal or glass plate by means of a ruling engine which is a tedious and expensive process. The gratings can be replicated by using a plastic material applied to the surface of the grating which takes the pattern of the grating and reproduces it in the plastic. This is usually done with a curable resin solution and at low temperature to avoid damage to the grating. Using this technique, it is possible to make low cost gratings which

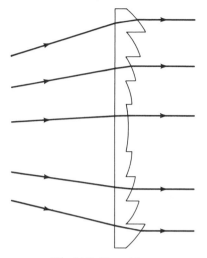

Fig. 16-7. Fresnel lens.

Fig. 16-8. This wide angle Fresnel window lens expands the view when mounted on a window-pane. The disc is a flat sheet with grooved lens on one side. (*Courtesy Optical Sciences Group*)

can be used for light analysis or for displays, or even to make an interesting form of iridescent jewelry.

There are a number of applications for plastics in optics which involve the interaction with polarized light. Polarized light is distinguished from ordinary light which is called *incoherent light*. Incoherent light has wavelengths varying over a range of values, the phase of the individual wave trains do not have any phase relationship with each other, and the plane of the electric and magnetic vectors of the individual wave trains have random orientation with respect to each other. Looking into a beam of incoherent light, the electric vector can have any angle as shown in Fig. 16-9. When the

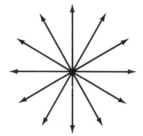

Fig. 16-9. End view of electrical vectors incoherent light.

light is plane polarized, all of the light which passes through the polarizing device has the electric vectors parallel and the magnetic vectors parallel to each other. The effect of the polarizer can be likened to a picket fence through which you can attempt to make wave trains with a rope or string. The only waves that will come through are those with the plane of vibration parallel to the fence pickets as shown in Fig. 16-10.

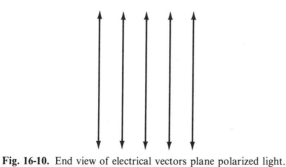

Fig. 16-10. End view of electrical vectors plane polarized light.

There are several ways in which light can be plane polarized. In some light sources because of the effect of strong electric and/or magnetic field present when the light is generated, it is polarized (A). When light is reflected at low angles from a dielectric medium which is transparent, the reflected light is polarized parallel to the surface and the transmitted light is polarized perpendicular to the surface (C). Naturally birefringent (double refracting) materials can be made into prisms which pass light polarized in one plane and reflect out of the prism light polarized in the plane at right angles (D). The method most generally used for generating polarized light is by passing incoherent light through a polarizing filter having the property of absorbing light which is polarized in the principal absorbing plane of the material and passing through polarized light whose electric vector plane is perpendicular to the absorbing plane (B). The various methods of producing polarized light are shown in Fig. 16-11.

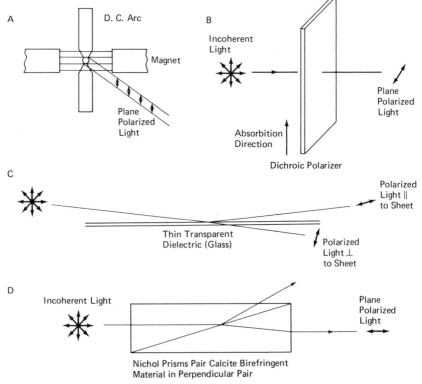

Fig. 16-11. Methods of generating polarized light.

The dichroic polarizer widely utilized to produce polarized light employs polymers in the dichroic materials with the polarizing capability. Figure 16-12 shows schematically the structure of a typical material. It consists of a polymer with the molecules oriented strongly in the direction desired for polarization. The polymer has attached color absorbing structures along its length. When light passes across the polymer molecules perpendicular to the length of the molecule, there is a minimum of interaction with the color absorbing centers. When the light passes across the polymer chain parallel to the chain, there is a high degree of light absorption. With a substantial thickness of the oriented material, the light that passes through is plane polarized in the plane perpendicular to the orientation direction. Examples of such dichroic polarizers are polyvinyl alcohol, oriented, with iodine absorbed on the alcohol side groups and polyvinylene which is made from oriented polyvinyl alcohol by heating to a temperature which causes splitting off of water to form unsaturation along the polymer chain. Multiple conjugate unsaturation in an organic molecule produces a light absorbing structure. In addition to the two examples given, it is possible to make absorption type dichroic polarizers by attaching dichroic dyes on to oriented polymer chains which have an affinity for the dye. Most dichroic polarizers pass about 45% of the incident energy through as plane polarized light.

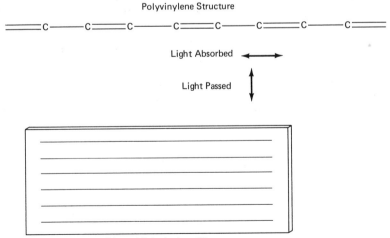

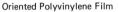

Fig. 16-12. Structure of a dicaroic polarizer.

One of the simplest and most interesting applications for polarizers in optical applications is the use of two polarizers to control the amount of light passing through the pair. If two polarizers are used serially on a light beam, the orientation of the second polarizer to the first will determine the amount of light passed. If the planes of polarization are parallel, then all of the light passed by the first polarizer will be passed by the second polarizer. In the planes of polarization are at right angles for the two polarizers then no light is passed. At angles between, differing amounts of light are passed based on the angular relation of the two polarizers to each other.

Two crossed polarizers are frequently used to inspect transparent materials placed between them for optical activity, either for birefringence or for optical rotary effects. Birefringence effects are produced by materials with a regular ordered structure that allows light to pass through at one orientation at a higher velocity than at another orientation. As a result of this, the two wave trains generated by the different velocities cause the phase angle of the light beam to change so that the light exits from the birefringent material with a different optical orientation than the entering light. Usually this is at an angle where part of the light can pass through the second polarizer. In many cases the emerging light is elliptically or circularly polarized instead of plane polarized so that part of the beam is absorbed in the second polarizer. Light which is elliptically or circularly polarized has the electric vector varying in value and angle along the wave train and it behaves similarly to incoherent light on passage through a plane polarizer.

Optical rotary effects are also referred to as optical activity and are caused by optically assymetric groups in the structure of the material. A typical optically active group would be a carbon atom which has a different organic group attached at each position on the molecule. Cellulose and sugar have this type of structure. In this case the beam of polarized light has its plane rotated as it passes through the structure and the light emerges as plane polarized light with a different plane of polarization. By rotating the second polarizer's plane of polarization it is possible to find the exit angle and pass all of the light.

The crossed polarizer effects of both types are used in analysis work. The concentration of optically active organic materials is determined by the degree of rotation. In plastic processing the residual strains in molded materials as well as the degree of orientation of polymers is determined by the effect on polarized light. The frontispiece in this book, which is a visual display of stress in a molded part, is done by the use of crossed polarizers. Crossed polarizers are used with special wave plates to control the amount of light that passes through an optical system.

Another application for the crossed polarizers is in electrically modulating the strength of a light beam. Electric fields have the effect of making certain substances variably birefringent. Currently the most important of these materials are the liquid crystal materials mentioned in Chapter 15. Many liquid crystals are low polymers

Fig. 16-13. The coated portion of this plastics lens soaks up moisture to retain clear vision. (*Courtesy National Hydron Corp.*)

with highly polar side chains. By varying the field on the material the birefringence is varied and the light transmission is controlled. This effect is used in optical devices and has application in communications systems, especially those using lasers as a signal source.

Plastics materials such as methyl methacrylate are currently used to make one type of laser material. When the appropriate luminescent dyes are incorporated into the material, the acrylics can be made into large laser units. One advantage of using acrylics is that very large clear castings can be made. As a result, large amounts of laser light can be produced at low volume densities of light. Consequently, heating effects are at a minimum. Optical systems can be used to concentrate the light from the large laser elements.

Plastics are suitable for most optical applications that utilize transparent materials, including color carriers. Color filters have all types of standard transmission characteristics can be made and, because of the uniqueness of the polymer structure, a large number of dichroic and trichroic materials are possible that have different colors when viewed from different angles. As previously mentioned, one application for this is in polarizing filters. An interesting application is in sunglasses where the tinting effect is combined with the polarizing effect to get sunglasses that are particularly effective against low angle glare. The lens materials are polarized in the vertical plane

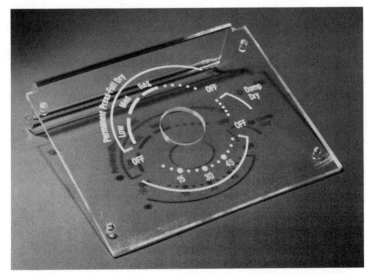

Fig. 16-14. This illuminated panel for a dryer is molded of acrylic resin.

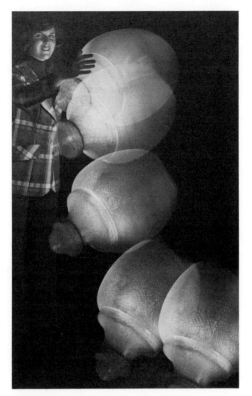

Fig. 16-15. Street lighting globes are blown from molded of polycarbonate to resist malicious damage. (*Courtesy General Electric Co.*)

and the low angle ground reflections are polarized in the horizontal plane so that the glare light is strongly absorbed in the lenses. An antifogging coating on a plastic lens soaks up condensed moisture (Fig. 16-13).

Within the limitations on the physical properties which generally restrict plastics to low precision optics, plastics materials have found wide applications in optical products that range from lights to binders for electroluminescent phosphors to fiber optics and lasers. They represent an easily worked material with a wide range of desirable optical properties. In this chapter the discussion has been limited to the differences between plastics and other optical materials and to some of the unique design possibilities that are especially good for plastics. For detailed optical design, the designer is referred to the optical design literature where the principles of optics apply as well to plastics as to other transparent materials. Using the optical arts and the special properties of plastics, unique products are attainable. The breakage resistance of the plastics has placed them into many common functional applications such as instrument windows (Fig. 16-14) and street lighting globes (Fig. 16-15).

Chapter 17

Other Design Applications for Plastics

This chapter includes some of the newest and most interesting fields of application for plastics and the design efforts made in these areas. Plastics are used uniquely in these applications because of specific material properties and material modifications that can only be done with polymer-based materials. By incorporating some innate characteristics of the materials and adapting them to operate in unusual environments, some of the most significant problems of mankind are being solved by the use of plastics.

One of the most interesting areas of application is in biological systems. In the introduction to the book it was pointed out that natural polymers are one of nature's oldest materials. New polymers are in use to help nature in biological systems, particularly in medicine. The applications in medicine are a good starting point for discussion since they include both mechanical and chemical applications and show the makeup of the field we could call *bioplastics*.

The heart valve (Fig. 17-1) which is often used in surgery to correct heart deficiencies was a spectacular contribution to medicine. In order for it to be successful it required first, ingenuity in designing a part that would function as a replacement for the mitral valve and to perform as well as the one replaced long enough to justify the risk involved in the operation. Second, it also required using a material that would function in the highly complex environment of the human circulatory system without being degraded and without causing harm to the circulatory system. The use of special silicone rubber materials designed expressly for the purpose and tested for years to determine possible adverse in-vivo effects combined with good design resulted in a mechanical device that has been a major lifesaver. The role of the plastics designer in an area such as this is one of knowing material limitations, processing problems and of

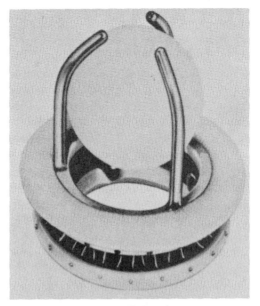

Fig. 17-1. Heart valve.

devising test procedures to monitor the performance of the part in the patient as well as in continuous laboratory testing. The services of skilled physicians in determining patient reaction, of biochemists to supply information on tissue reaction and tissue compatibility, and of chemists to devise ways in which the material can be modified (for example, by reacting a coating of heparin to the surface of the part to reduce clotting dangers) are part of a design effort of this type. The success with the heart valve has led to a great deal of work with surgical implants that are essentially plastics repair parts for worn out parts of the body. It is possible to conceive of major replacements of an entire organ such as a kidney or a heart by combining the plastics skills with tissue regeneration efforts which may extend life indefinitely. Depicted in Fig. 17-2 is a pulse generator which has several plastics components. This is used to time the heart action.

While it would be difficult to enumerate all of the efforts in the area of implants, some of the significant ones are: (1) the implanted pacemaker, (2) the surgical prosthesis devices to replace lost limbs, (3) the use of plastic tubing to support damaged blood vessels, and (4) the work with the portable artificial kidney. The kidney application

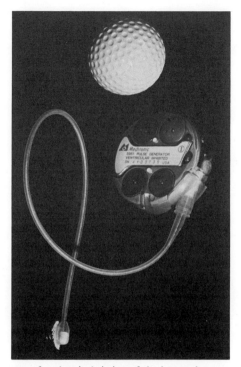

Fig. 17-2. Pulse generator for electrical timing of the human heart employs silicone plastics. (*Courtesy Medronic Co.*)

illustrates an area where more than the mechanical characteristics of the polymer are used. The kidney machine (Fig. 17-3) consists of large areas of a semipermeable membrane, a cellulosic material in some machines, where the kidney toxins are removed from the body fluids by dialysis based on the semipermeable characteristics of the plastics membrane. A number of other polymer materials are under study for use in this area, but the basic unit is a device to circulate the body fluid through the dialysis device to separate toxic substances from the blood. The mechanical aspects of the problem are minor but do involve supports for the large amount of membrane required.

The construction of full dentures was one of the earliest successful applications which involved stringent requirements. The oral environment is corrosive and abrasive. A denture is subjected to high stresses and the materials utilized must be nontoxic, nonirritant to tissues, and resistant to staining. These severe demands and long-term stabil-

Fig. 17-3. The GE DuaLung employs a plastics membrane. Interposed between the sweep gas and blood is a membrane which prevents the blood from interfacing with the gas. (*Courtesy General Electric Medical Systems Division*)

ity have been satisfactorily met by methyl methacrylate polymers and copolymers. Other dental applications include the recent use of composite materials to replace amalgam filling materials, and the use of plastics instead of metal braces in orthodonture work.

The major area of application for plastics in bioscience is in the two areas indicated. The plastics make interesting materials to be used for mechanical implants into all living systems, including animals and plants where they can serve as repair parts or as modifications of the system. Figure 17-4 shows a variety of plastics used in hospitals. The other applications are based on the membrane qualities of plastics which can control such things as the chemical constituents that pass from one part of a system to another, the electrical surface potential in a system, the surface catalytic effect on a system, and in some cases the reaction to specific influences such as toxins or strong radiation. A Dacron artery is shown in Fig. 17-5.

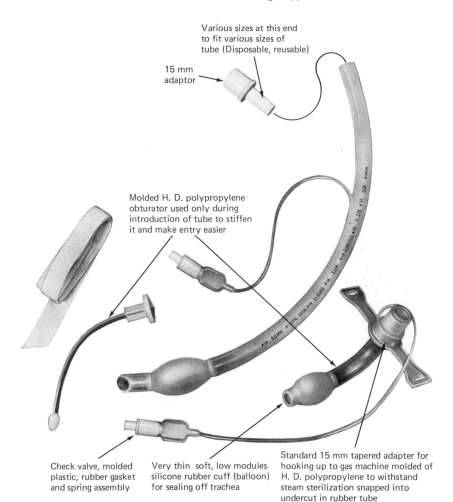

Various sizes at this end
to fit various sizes of
tube (Disposable, reusable)

15 mm
adaptor

Molded H. D. polypropylene
obturator used only during
introduction of tube to stiffen
it and make entry easier

Check valve, molded
plastic, rubber gasket
and spring assembly

Very thin soft, low modules
silicone rubber cuff (balloon)
for sealing off trachea

Standard 15 mm tapered adapter for
hooking up to gas machine molded of
H. D. polypropylene to withstand
steam sterilization snapped into
undercut in rubber tube

Fig. 17-4. Tracheal intubations can be life saving. Most intubations are as long as the operative procedures; some critical airway needs last hours or days. (*Courtesy Medical Engineering Corp.*)

Plastics have a unique contribution to make in exploring new environments. Their applications in space vehicles are well known and are generally of a mechanical nature. Plastics are used extensively in gear for exploring under the sea. One unique application for plastics now used in the sea environment but which may well be important in space is in filtering gases. This specific application involves the use of membranes made from special silicone compounds which

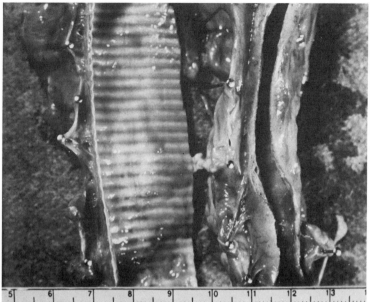

Fig. 17–5. Open view of healed 40-needle-per-inch Arterial Prosthesis of Dacron (left) 27 months after implantation into the iliac tree of a patient who died from unrelated disease. It is compared with the patient's own iliac arteries (right).
This is the first type of artificial artery which shows the presence of an elastomechanical layer surrounding the healed man-made artery.
This photograph demonstrates the similarity of the elastomechanical layer of the prosthesis to the normal areolar dissection plane around the human arteries. (*Courtesy Surgical Uses of Plastics*, S. A. Wesolowski *et al*. Mercy Hospital Rockville Center, N.Y.)

permit dissolved oxygen in seawater to permeate the membrane in one direction and to allow carbon dioxide and carbon monoxide to pass through in the opposite direction. A large membrane pack of this type will act like an artificial gill, permitting a swimmer to breathe like a fish and remain submerged for much longer periods of time than are possible with scuba equipment. Speculative fiction has man returning to live in the seas, and this type of application may make it possible. Their application in spacecraft is obvious as a part of a continuously recycled air support system.

The oxygen permeability of silicone materials is just one example of the selective permeability of polymers. The application in the artificial gill is a dramatic one, and the effect is used in many other areas. The selection of materials to package foods is based on the permeability of the materials to oxygen, water vapor, and, in the case of packaging bananas, to ethylene gas which is used to artificially ripen the bananas. Selective permeability provides chemical separations, one of the most interesting of which is the use of TFE materials to separate the hexafluorides of the different isotopes of uranium. There are a number of industrial gas separation systems that use the selective permeability of plastics to separate the constituents. In design problems relating to such applications, the designer must consider the environmental conditions to determine whether the materials having the desired properties will withstand the temperatures and physical and chemical stresses of the application. Frequently the application will call for elevated temperatures and pressures. In the case of uranium separation, the extreme corrosivity of the fluorine compounds precluded the use of any material but TFE. The TFE material, however, required careful design to make a sturdy membrane because of the poor mechanical properties of the TFE resins.

One of the major problems facing our civilization is the availability of pure water. The largest source of water located near many cities is the ocean, but the ocean is filled with large amounts of dissolved salts. To recover water from the sea by any of the conventional distillation processes is extremely wasteful of energy. Plastics membranes are being used in a system that could well pave the way to large-scale water recovery from the sea. The process is reverse osmosis. When water is separated from a concentrated solution of a salt by a semipermeable membrane, there is a pressure which drives the pure water into the solution for dilution. The driving force is the concentration gradient and it is in the form of a pressure which is related to the difference in the vapor pressure of the water and the vapor pressure of the solution at the temperature at which the process takes place. By applying a pressure greater than the osmotic pressure to a solution, the direction of flow of the water is reversed and pure water is removed from the solution. This is done by the use of a water permeable plastics membrane held deep enough under the sea so that the hydrostatic pressure is greater than the osmotic pressure of the seawater. The water distills out of the solution through the membrane and is pumped to the surface. Large areas of the membranes, mechanically supported to withstand the very high pressures are essential to make the process perform rapidly for economical production. Cellulosic plastics are used for the membrane, but any water

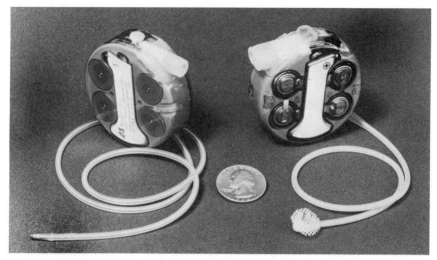

Fig. 17-6. Heart timing unit.

vapor permeable material is a good possibility, provided the film has good mechanical properties. Designing the membrane structure for a reverse osmosis plant is a difficult project, particularly in view of the fact that in addition to the pressure exposure, the presence of strong concentrations of dissolved minerals is a hostile environment for plastics. There have been several very ingenious designs for membrane structures using naturally strong shapes such as arches and tubes reduced to a scale where the amount of surface area for diffusion per unit volume is very high. Other innovations in design and fabrication of these large area membrane structures could easily lead to a significant breakthrough in the availability of an unlimited supply of pure water.

Plastics have had a long history of use in water problems using a special type of polymer material. Ion exchange resins have been used for many years to remove dissolved materials from water and to make high purity water. The quality of ion exchange treated water is probably better than water purified by any other means. The ion exchange resins are polyelectrolytes. These resins are polymers that contain either acid or alkaline side groups capable of reacting with dissolved ions in the water. A standard system used an acid substituent that forms an insoluble salt with ions such as calcium and strontium and removes them from the water and replaces the mineral ion with hydrogen. The ion exchange resins can be regenerated by passing another solution through the bed such as a mild mineral acid which removes the attached ions and replaces them with hydrogen again so that the resin can be reused.

Polyelectrolytes such as the ion exchange resins form an interesting group of materials because of their ability to interact with water solutions. They have been used in medical applications involving the removal of heavy metal ions from the human body. They can be used to interact with external electric fields and change their physical properties drastically as is illustrated by the fact that some electrically active liquid crystals are polyelectrolytes of low molecular weight. Another application for polyelectrolyte materials is in the forming resins with unusual physical properties with regard to adhesion. The incorporation of small amounts of organic acid materials into poly-olefin structures results in materials that have excellent adhesion to metals, paper, glass, and a variety of other materials. In addition, the materials with electrolytic structures can have a metal ion incorporated into the structure resulting in the formation of the ionomer type of resin that has much better mechanical properties than the basic polymer material.

Chemically active polymers such as the polyelectrolytes have been used to make artificial muscle materials. This is an unusual type of mechanical power device that creates motion by the lengthening and shortening of fibers made from a chemically active polymer by changing the composition of the surrounding liquid medium, either directly or by the use of electrolytic chemical action. Obviously this form of mechanical power generation is no competitor to thermal energy sources, but it is potentially valuable in detector equipment that would be sensitive to the changing composition of a water stream or other environmental flow situation. By using direct mechanical action from the artificial muscle, it would be possible to produce reliable sensing and control devices without electrical and electronic equipment. Another interesting application would be to drive prosthetic devices where the action would be similar to the muscle reaction in the body. This unusual type of chemically induced motion should be an interesting one to explore for the solution of unusual problems where conventional approaches do not work.

One of the problems that faces our civilization is the fact that the pressure on natural resources is hindering progress. At present an energy crisis exists which has led to a materials crisis in plastics. Petroleum is currently the source of raw materials for most high volume polymers. Oil and substitute resources such as coal are in limited supply, and it may well be that another approach to the problem is required. Ingenuity in the applications of materials, the province of the designer, and the use of materials which seem to be uniquely modifiable, such as plastics, are needed. We must develop renewable resources for the plastics. We can take as an example papermaking which now consumes our forests at a rate which is difficult to maintain. Unlike the uses for wood, which are generally long-term use goods, most paper is used for newspapers and periodicals which are read and discarded, loading our solid waste disposal system and adding mountains to our trash. Plastic papers have been developed as substitutes for cellulose paper, but the economics are poor since the resins are costly and must be converted to paper fiber. We shall show, by using some of the intrinsic characteristics of plastics, that it is possible to save the forests and at the same time to improve the information distribution system represented by newspapers and periodicals without changing the reading habits of people.

The justification to go to a higher cost paper material would be part of a system where the paper is continuously reused. The sole function of the paper is to carry printed information. Usually the information is carried by truck and other carrier to the reader. By using a local printing terminal, located either at your newsstand or in your home, delivery can be made by wire. By using a material which has erasable printing generated by the remote printing terminal, the newspaper or periodical could be printed at its destination. Plastics materials can be used to make erasable printing media by a number of different techniques. Photo changing dyes could be incorporated into the structure of the polymer. The printer could change the dye to the colored form to read, and the material can be bleached with another unit that would reverse the photo coloring process. An ionic type resin can be incorporated into the plastics and used to color the printed area by the use of an indicator type reaction with an organic acid or base. Another method would be to use a thermal printer in conjunction with liquid crystal type materials which would alter the state of the liquid crystals in the printed areas. Applying heat and electrical fields to the printed sheet would erase the printing. Other schemes involving dichroic dyes with heat and electrical fields are also possible. Each of the possibilities could use either the polymer structure of the plastic substrates or its durability, or both.

This approach would recycle the material for carrying the printed messages at the point of use, eliminating handling and distribution costs, and would require a fraction of the enormous amount of paper now consumed in delivering news and other literary material. The newspaper or periodical would have the familiar size and appearance and would present little change to the reader. The convenience of real on time home delivery and other built in aspects of the system would make it a useful successor to the present one.

The design opportunities for plastics materials in the future will be in such areas as this where the special properties of the material can be made to accomplish a unique result. Ingenuity in the application of materials has been the thrust of the plastics industry, and it will present new opportunities in the future.

Several other interesting material developments that may be useful to the plastics designer in the future will be mentioned. Since one key step in any design is the material selection, an important aspect of the designer's responsibility is to be familiar with the range of material possibilities. Plastics materials interact with high energy radiation by giving bursts of visible light. This can be made selective for particular types of radiation and the effect has been used to make materials for scintillation counters to measure gamma radiation and particle streams such as alpha particles and beta rays. The ability of some resin systems to do this may be useful in schemes for handling the radiation output of nuclear devices, including the radiation from the fusion power machines under development. Obviously the application is not for shielding, which the heavy metals do much better, but rather for an energy level reduction system that would convert the high energy radiation to forms which would be more useful in power distribution.

There has been a great deal of interesting work done recently in attaching active enzyme materials to plastics substrates to convert simple organic molecules into the more complex forms used in biological processes. This technique makes available a catalyst bed capable of doing large-scale synthesis of materials such as proteins and carbohydrates that are essential to life processes. With a major food crisis looming as a result of the rapidly increasing population of the world, it may be necessary to revive the possibility of synthetic food production. Since farm land is being depleted and recurring drought conditions reduce food supplies, it is likely that synthetic food will be a necessary supplement. The selective action of enzyme membranes may be a way to approach the food synthesis problem.

In this chapter we have covered some interesting applications of plastic materials that are more potential than current. The purpose is to indicate the versatility of the polymer-based materials which can be varied to perform special functions. Nature has used these materials to build the structures of life forms with a great deal of success. By applying ingenuity to these amazingly adaptable materials, we can produce structures which add to the survival capability of life under severe environments and improve the quality of life under normal environments. The designer's role in fitting the possibilities to the needs is one that is increasingly important and it is hoped that this book may be a useful tool in the effort. . . .

Index

AC effects, 22, 23
ADC, 301
adherent, 43
agitator, 212
air supported structures, 138, 141
Alfrey Jr., Turner, 34, 35
allyl diglycol carbonate (ADC), 301
Anchor Plastics Co., 165
annealing, 198
Arrhenius Relationship, 34
arterial prosthesis, 322
artificial gill, 322
ASTM, 275
audio circuits, 285
autoclave molding, 189

Baer, Eric, 31, 34, 39, 69
Bell helicopter, 101
beams, 71
bearings, 114
Benjamin, B. S., 30, 134, 136, 139, 143, 144, 145
benzene ring, 12
Bernard, J., 198
biological systems, 317
bioplastics, 317
bioscience, 320
birefringent, 25, 300, 310
blow molding, 168
Boltzmann, 29, 34, 35, 37
borescope, 306
boron epoxy, 100
bosses, 158
brittle transition, 20
butter test, 216

carbon filament, 138
casting, 181

cast plastics, 207
catalytic, 43
charge carrier migration, 21
chemical effect, 289
Chemical Engineers Handbook, 61
coaxial cables, 282
cold flow, 8
collimation, 303
color, 300
computer flow chart, 269
computer sequence, 84
conoid, 136
contact lenses, 301
contact molding, 188
corrugated section, 234
corrugated sheets, 129
coupling agents, 56, 57
coupling currents, 279, 280
creep, 8, 86, 204, 237
crazing, 206
crossed polarizer, 312
crosslinked, 10, 11
crystalline melt temperature, 19, 22
crystalline solids, 4
crystalline structures, 3
crystalline transition, 19
crystallinity, 202
cycling loading, 90, 93
cycling stress, 38, 93
cystoscope, 306

damping designs, 98
debugging, 212
deflection time, 67
deformation fracture, 60
Dentique, 170
design procedure, 224
design steps, 230

329